Wael A. Altabey
Composite Materials and Structures

Also of interest

Hybrid Composite Materials and Manufacturing.
Fibers, Nano-Fillers and Integrated Additive Processes
Vipin Kumar (Ed.), 2024
ISBN 978-3-11-101934-5, e-ISBN (PDF) 978-3-11-101954-3,
e-ISBN (EPUB) 978-3-11-102013-6

4D Printing of Composites
Suong V. Hoa, 2024
ISBN 978-3-11-066844-5, e-ISBN (PDF) 978-3-11-066854-4,
e-ISBN (EPUB) 978-3-11-066862-9

Polymer Matrix Composite Materials.
Structural and Functional Applications
Debdatta Ratna, Bikash Chandra Chakraborty, 2023
ISBN 978-3-11-078148-9, e-ISBN (PDF) 978-3-11-078157-1,
e-ISBN (EPUB) 978-3-11-078170-0

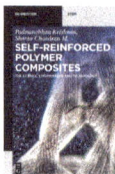

Self-Reinforced Polymer Composites.
The Science, Engineering and Technology
Padmanabhan Krishnan, Sharan Chandran M, 2022
ISBN 978-3-11-064729-7, e-ISBN (PDF) 978-3-11-064733-4,
e-ISBN (EPUB) 978-3-11-064740-2

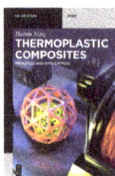

Thermoplastic Composites.
Principles and Applications
Haibin Ning, 2021
ISBN 978-1-5015-1903-1, e-ISBN (PDF) 978-1-5015-1905-5,
e-ISBN (EPUB) 978-1-5015-1161-5

Wael A. Altabey

Composite Materials and Structures

—

Artificial Intelligence-Based Structural Health Monitoring

DE GRUYTER

Author
Prof. Wael A. Altabey
Department of Mechanical Engineering
Faculty of Engineering
Alexandria University
Alexandria (21544)
Egypt
Wael.altabey@gmail.com

ISBN 978-3-11-914546-6
e-ISBN (PDF) 978-3-11-221309-4
e-ISBN (EPUB) 978-3-11-221340-7

Library of Congress Control Number: 2025940749

Bibliographic information published by the Deutsche Nationalbibliothek
The Deutsche Nationalbibliothek lists this publication in the Deutsche Nationalbibliografie;
detailed bibliographic data are available on the Internet at http://dnb.dnb.de.

© 2025 Walter de Gruyter GmbH, Berlin/Boston, Genthiner Straße 13, 10785 Berlin
Cover image: Wael A. Altabey
Typesetting: Integra Software Services Pvt. Ltd.

www.degruyterbrill.com
Questions about General Product Safety Regulation:
productsafety@degruyterbrill.com

Preface

This book provides a comprehensive and easy structural health monitoring (SHM) course in composite structures. It is a good technical resource for supporting engineers and researchers at both undergraduate and graduate levels, and also practicing engineers. The scope is not limited to any particular engineering discipline. This book also intends to provide engineers and researchers with an intuitive understanding of composite materials, manufacturing processes, theories and failure analysis methods, failure mechanisms of composite materials, fatigue behavior of composite materials, and SHM of composite structures, including the capability of nondestructive testing (NDT) techniques for SHM, and the principle of the artificial intelligence algorithms in SHM in composite structures such as machine learning, deep learning, artificial neural networks. This book is the first book for learning the SHM of composite structures and NDT techniques by easier, visual, and applied methods and connects between course examples and practical applications. Moreover, this is the first book to have a complete chapter for applying the SHM methods in a group of case studies, such as composite pipelines and plates, using various NDT methods integrated with artificial intelligence algorithms. These interactive case studies and solutions serve as effective, self-testing, as well as excellent preparation methods for SHM projects. In addition, this is the first book to use artificial intelligence algorithms in SHM in composite structures such as machine learning, deep learning, and artificial neural networks. The visual attractiveness of the book is enhanced by numerous new full-color graphic illustrations. It is appreciated for its readability, while being recognized for its technical strength and comprehensive coverage of SHM of composite structures and NDT techniques. The objectives of this book are to:

1) incorporate knowledge learned in composite materials, basic concepts and terminology, composite manufacturing, fabrication, and processing;
2) incorporate knowledge learned in the mechanical behavior of composite materials;
3) incorporate knowledge obtained from SHM in composite structure theory and the differences between them;
4) reinforce competence in using the artificial intelligence algorithms for SHM in composite structures and then to determine high performance in damage identification;
5) obtain a working knowledge in the use of proper SHM in composite structures under steady and variable loadings;
6) apply the principles of the finite element method in SHM in composite structures;
7) use the principles of artificial intelligence algorithms in SHM in composite structures; and
8) use the basis of NDT, various methods integrated with artificial intelligence algorithms in SHM of composite structures, with case studies, such as composite pipelines and plates, through analytical and numerical approaches using ANSYS and MATLAB built-in algorithms.

https://doi.org/10.1515/9783112213094-202

Contents

Chapter 3
Design concepts for composite materials/structures —— 24

Chapter 4
Composite manufacturing, fabrication, and processing —— 32

Chapter 7
Case studies on structural health monitoring of composite structures —— 103

Chapter 1
Introduction to composite materials

Recently, composite materials have been used in engineering fields such as aerospace, mechanical, and structural fields, and also in the construction industry, shipbuilding and submarines, automotive, and nuclear and chemical industries. The importance of them to use in these fields is because they provide high strength and stiffness per weight ratios, also they provide fatigue and corrosion resistance and damage tolerance capacity; hence, they are a good viable alternative to metallic materials.

The increasing application of composite materials in various engineering fields, especially in aerospace, mechanical, and structural systems, has imposed more challenges. Safety, serviceability, sustainability, and reliability are critical requirements for using composite materials in both new and existing systems. Furthermore, this has driven the development of methodologies for real-time monitoring, evaluation, and damage identification in composite materials [1, 2].

There are several types of composite material damage such as fiber breakage, matrix cracking, debonding between the matrix and the fibers, and delamination or interlayer cracks. These types are key factors in the failure of composite materials because such damage can significantly reduce strength and structural safety. Structural health monitoring methods serve as a guide to preventing catastrophic failure in these types of structures. Also the detection system will collect important data regarding the evolution of structural damage, which can be used to create maintenance or repair schedule to ensure the structure's safety.

Structural health monitoring aims to assess the condition of structures by analyzing data collected from an active sensory network system distributed across the structures to permanently monitor and thus decide the presence or absence of damage. The primary objectives of a structural health monitoring system are to ensure the integrity, serviceability, and sustainability of structures through the development of automated systems for damage detection, inspection, and continuous monitoring. In recent years, numerous studies have described a wide range of quantitative data analysis techniques that have been developed and successfully applied for damage detection in both metallic and nonmetallic structures, such as composite materials, using nondestructive testing methods [3–5].

The Lamb wave technique is one of the most utilized techniques and has a significant ability to detect different types of damage in composite materials in a faster and more cost-effective way. However, this approach requires high expertise for monitoring complex structures. Additionally, the Lamb wave technique is characterized by its ability to inspect large structures using a small number of transducers. However, determining the optimal positions for the transducers remains a critical challenge.

The piezoelectric transducers are the most commonly used transducers with the guided Lamb wave approach due to their advantages in terms of weight and cost

https://doi.org/10.1515/9783112213094-001

[6, 7]. Lamb waves detect damage based on wave propagation across the structure's span. These waves interact with the damage, and baseline signals, representing the non-damaged state, are used for comparison with the damaged signals to observe changes in response to the Lamb waves.

The other important advantage of the Lamb wave method is that the wave can travel across long distances with high sensitivity to interference along the propagation path, even in materials with high attenuation, such as composites [8], which makes it possible to easily test large areas, such as the composite wing skin of an airplane. Lamb waves are also capable of detecting not only surface damage but also internal damage, as they can probe the full thickness of the material by utilizing different modes of Lamb waves. Overall, damage detection methods based on Lamb waves offer the following benefits: (1) the ability to detect large structures without disturbing the coating or insulation of the structure being tested; (2) full detection of the cross section of a structure over long lengths; (3) the capability to detect different types of damage in a more cost-effective manner; and (4) operation with very low energy consumption and cost [8].

This chapter presents a comprehensive systematic review of the scientific research and development of structural health monitoring and damage identification in composite structures under fatigue loading using the Lamb wave technique. First, it is essential to understand the mechanical behavior of composite materials, including failure mechanisms, fatigue behavior, fatigue load effect of composite materials, and the failure criteria under fatigue loading. Following this, a review of the most widely used nondestructive testing methods based on structural health monitoring for detecting fatigue damage in composite materials should be conducted. The ability of Lamb waves to detect fatigue damage in composite materials is discussed to provide insight into the different damage mechanisms that occur due to fatigue loading. Determining the optimal number and positioning of piezoelectric transducers on the surface of composite materials is one of the critical challenges addressed in this work. The application of artificial intelligence algorithms for predicting fatigue damage in composite materials will be discussed. This is one of the most desirable methods to demonstrate how to combine artificial intelligence-based methodologies with Lamb wave application in predicting fatigue damage through a training process.

The chapter is organized as follows: the next section outlines the composite structures, including their applications, characteristics, failure mechanisms, fatigue behavior, and failure criteria under fatigue loading. In the next section, the structural health monitoring techniques are discussed, and further divided into sections to introduce the capability of nondestructive testing techniques and the characteristics of piezoelectric sensors for structural health monitoring, as a foundation for the main technique will be reviewed in this chapter, precisely the Lamb wave-based structural health monitoring of composite structures. In the context of this method, several sections delve into the technique's applications, characteristics, history, literature, modeling, and simulation. The selection of the optimal number and positions of piezoelectric transducers are key points discussed in this chapter. Also the Lamb wave-based damage identification principle is simplified and explained in detail, making it accessible to any researcher beginning their academic career in this field. In the next section, structural health monitoring is linked with modern technology and artificial intelligence algorithms, where the application of artificial intelligence algorithms, such as machine learning, deep learning, and artificial neural networks, to enhance the performance of structural health monitoring techniques for damage detection in composite structures is thoroughly discussed. In the final section of this comprehensive systematic review, the number of reviewed publications per time period is illustrated in a statistical figure. The chapter concludes by summarizing the findings of the review and identifying niche research areas that require further exploration.

1.1 Introduction to materials

Material is the fundamental element of all natural and synthetic structures. Figuratively speaking, it embodies the structural concept. Technological progress is linked to the continuous improvement of existing material properties, as well as the expansion of structural material classes and types. Typically, new materials emerge from the necessity to enhance structural efficiency and performance. However, new materials, in turn, often provide opportunities to develop advanced structures and technologies, which subsequently present materials science with new challenges and tasks. One of the most notable examples of this interconnected process in the development of materials, structures, and technology is observed in composite materials.

1.1.1 Classification of materials

The classification of materials is shown in Figure 1.1.

1.2 Introduction to composite materials

1.2.1 What are composites?

Composite materials consist of two or more materials, which together produce desirable properties that cannot be achieved with any of the constituents alone. Fiber-reinforced composite materials, for example, contain high-strength and high-modulus fibers in a matrix material. Reinforced steel bars embedded in concrete provide an example of fiber-reinforced composites. In these composites, fibers are the principal load-carrying members, and the matrix material keeps the fibers together, acts as a load-transfer medium between fibers, and protects fibers from being exposed to the environment (e.g., moisture and humidity).

It is well known that fibers are stiffer and stronger than the same material in bulk form, whereas matrix materials retain their usual bulk-form properties. Geometrically, fibers have a near crystal-sized diameter and a very high length-to-diameter ratio. Short fibers, known as whiskers, paradoxically exhibit better structural properties than long fibers. To gain a full understanding of the behavior of fibers, matrix materials, agents that are used to enhance bonding between fibers and the matrix, and other properties of fiber-reinforced materials, it is necessary to consider certain aspects of materials science. Since this study is entirely focused on the mechanical aspects and analytical methods of fiber-reinforced composite materials, no attempt is made here to present basic materials science concepts, such as molecular structure or the interatomic forces that hold the matter together. However, an abstract understanding of material behavior is beneficial.

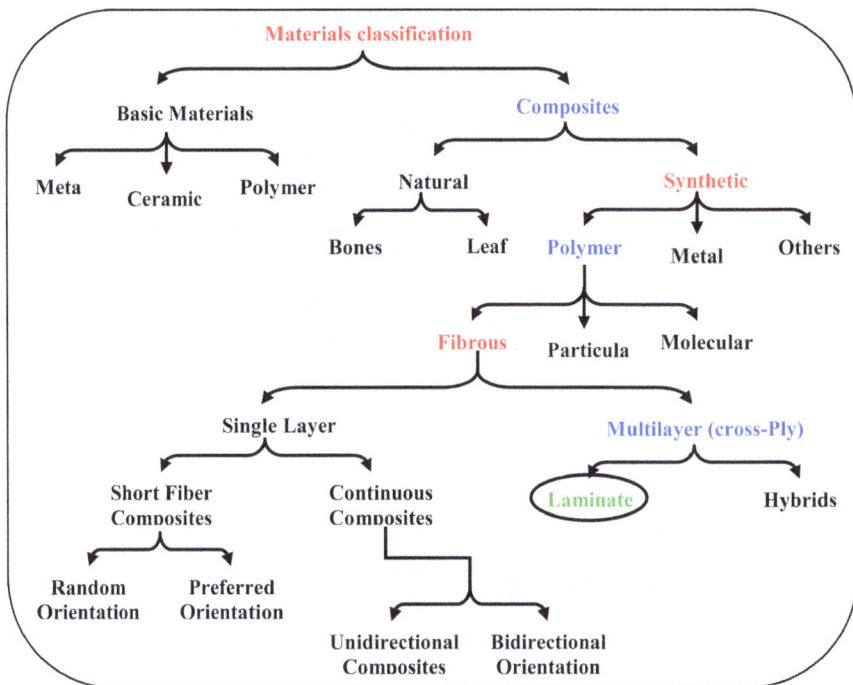

Figure 1.1: Classification of materials.

1.2.2 History of composite materials

Composites have been used for a very long time since 1500 BC. History contains some source of composite materials. An early example of composites is bricks made of clay reinforced with straw by Israelites. The reinforcing mud walls in houses with bamboo shoots, and glue-laminated wood was used by Egyptians in 1500 BC, and laminated metals were used in forging swords in AD 1800. It was found that rearranging wood could achieve greater strength and resistance to thermal expansion. Additionally, early natives of South and Central America demonstrated the use of composites by incorporating plant fibers into their pottery. This technique was employed to keep the clay from cracking during drying, rather than for structural reinforcement. In the 1930s, composites were modernized with the introduction of glass fiber-reinforced resins. These glass composites, commonly known as fiberglass, were applied in the construction of boats and aircraft. Since the 1970s, composites have been widely used due to the development of new fibers such as carbon, boron, and aramids, as well as new composite systems with matrices made of metals and ceramics. Some of the properties that can be improved by using composite materials include:

- strength,
- stiffness,
- corrosion resistance,
- weight,
- thermal insulation,
- wear resistance, and
- fatigue life.

1.2.3 Advantages and disadvantages

1.2.3.1 Advantages of composite materials
Summary of the advantages exhibited by composite materials, which are of significant use in the tub and vessel industry, is as follows:
1. Highly resistant to fatigue and corrosion degradation.
2. High "strength- or stiffness-to-weight" ratio. As mentioned above, weight savings are significant, ranging from 25% to 45% of the weight of conventional metallic designs.
3. Due to greater reliability, fewer inspections and structural repairs are required.
4. Directional tailoring capabilities are designed to meet specific design requirements. The fiber pattern can be arranged in a way that tailors the structure to efficiently withstand the applied loads.
5. Fiber-to-fiber redundant load path.
6. Improved dent resistance is typically achieved because composite panels are less prone to damage compared to thin-gauge sheet metals.
7. It is easier to achieve smooth, aerodynamic profiles for drag reduction. Complex, double-curvature parts with a smooth surface finish can be produced in a single manufacturing operation.

8. Composites offer improved torsional stiffness. This results in higher whirling speeds, a reduced number of intermediate bearings, and fewer supporting structural elements. Thus, the overall part count and manufacturing and assembly costs are reduced.
9. Highly resistant to impact damage.
10. Thermoplastics have rapid processing cycles, making them attractive for high-volume commercial applications that have traditionally been the domain of sheet metals. Moreover, thermoplastics can also be reshaped.
11. Like metals, thermoplastics have an indefinite shelf life.
12. Composites are dimensionally stable; that is, they have low thermal conductivity and a low coefficient of thermal expansion. Composite materials can be tailored to comply with a broad range of thermal expansion design requirements and to minimize thermal stresses.
13. Manufacturing and assembly are simplified because of part integration (joint/fastener reduction), thereby reducing costs.
14. The improved weatherability of composites in a marine environment as well as their corrosion resistance and durability reduce the downtime for maintenance.
15. Close tolerances can be achieved without machining.
16. Material usage is reduced because composite parts and structures are frequently manufactured to their final shape rather than being machined into the required configuration, as is common with metals.
17. Excellent heat sink properties of composites, especially carbon–carbon, combined with their lightweight nature, have expanded their use in aircraft brakes.
18. Improved friction and wear characteristics.
19. The ability to tailor the basic material properties of a laminate has enabled new approaches to the design of tubs and vessels.

The above advantages apply not only to tubs and vessels but also to common implements and equipment, such as air ducts, which have inherent damping and cause less fatigue and discomfort for the user.

1.2.3.2 Disadvantages of composite materials
Some of the associated disadvantages of advanced composites are as follows:
1. High costs of raw materials and fabrication.
2. Composites are more brittle than wrought metals and, therefore, more easily damaged.
3. Transverse properties might be weak.
4. The matrix is weak; therefore, it has low toughness.
5. Reuse and disposal may be difficult.
6. It is difficult to attach.
7. Repair introduces new problems for the following reasons:

- Materials require refrigerated transport and storage and have a limited shelf life.
- Hot curing is necessary in many cases that require special tooling.
- Hot or cold curing requires time.
8. Analysis can be challenging.
9. Matrix is susceptible to environmental degradation.

However, proper design and material selection can mitigate many of the aforementioned disadvantages.

1.2.4 Applications of composite materials

In recent years, composite materials have been used in several structural applications. Their usage has grown rapidly, and they are now widely used, becoming the dominant type of material in structural applications, as well as in the automotive, aerospace, marine, and sports equipment industries. This is due to their unique mechanical properties, especially their specific strength and stiffness, as well as their resistance to fatigue and ability to tolerate damage. For example, the new Boeing B787 has extensively embedded composite materials into its final aircraft structure, including the fuselage, which is entirely made of composite materials [9, 10], as shown in Figure 1.2.

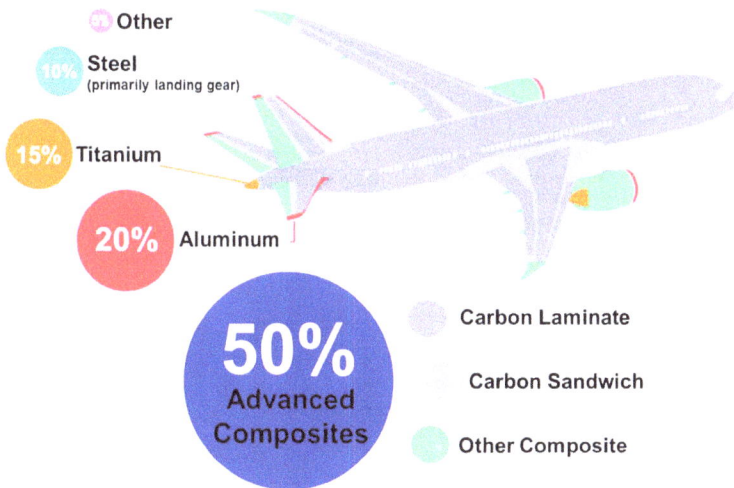

Figure 1.2: The percentage of the usage of composite materials in Boeing 787.

In recent decades, as composite materials are subjected to a combination of static and dynamic loads, such as fatigue, they have become one of the most challenging areas of research. It is known that around 80% of failures in mechanical and structural systems are related to fatigue [11].

1.2.4.1 Space applications

There are numerous applications of composites in various fields. They have been used widely in the fields of aircraft, space, automotive, sporting goods, and marine engineering. The use of fiber-reinforced polymers has experienced a steady growth in the aircraft industry, as weight reduction is critical for achieving higher speeds and increased payloads. By replacing conventional metal alloys with composite materials, the overall mass of the aircraft is reduced without decreasing the stiffness and strength of its components. Even though the costs of composite materials are higher, the reduction in the number of parts in an assembly and the savings in fuel costs make them more cost-effective.

Carbon fibers, either alone or in combination with Kevlar 49 fibers, have become the primary material in many wing, fuselage, and empennage components. For many military and commercial helicopters, fiber-reinforced epoxies are used in rotor blades. In numerous space vehicles, composites are utilized due to their dimensional stability over a wide temperature range. Many carbon fiber-reinforced epoxy laminates can be designed to achieve a coefficient of thermal expansion close to zero. Carbon fiber composites have much lower specific gravities, higher strength, and a higher stiffness-to-weight ratio. The applications of composites are shown in Figures 1.3 and 1.4.

Figure 1.3: Body of the plane made from composite materials.

Figure 1.4: Rotors made from fiber composites.

1.2.4.2 Automotive applications

The manufacturing and design of fiber-reinforced composite materials for automotive applications differ from aircraft applications. In terms of production volume, the number of automotive components required may range from 100 to 200 pieces per hour, compared to a few hundred pieces per year for aircraft components. Exterior body components, such as hoods or door panels, require high stiffness and damage tolerance (dent resistance), as well as a class A surface finish for appearance. Acceptable damage tolerance is achieved by using flexible resins, such as polyurethane. In engine components, the application of fiber-reinforced composites is still in the developmental stage, as fatigue loads at very high temperatures remain a challenge in these applications. In this regard, the development of high-temperature polymers and metal matrix composites could significantly improve the potential for composite usage in this field. Another advantage of many composite materials is their high internal damping. This property allows for better vibration energy absorption within the material, thereby reducing the transmission of noise and vibrations to neighboring structures. This high damping capacity can be particularly beneficial in many automotive applications.

1.2.4.3 Sporting industry applications

Fiber-reinforced polymeric composites are essential materials in the sports industry. Sporting goods such as tennis rackets, golf club shafts, fishing rods, bicycle frames, hockey sticks, and arrows are using fiber-reinforced polymers to reduce weight, dampen vibrations, and offer design flexibility. Weight reduction is achieved by substituting carbon fiber-reinforced epoxies for metal, which has led to higher speeds and quicker adjustments in competitive sports such as bicycle races or canoe races. In some applications, such as tennis rackets or snow skis, carbon or boron fiber-reinforced epoxies are used as the skin material, while a soft, lighter-weight urethane foam core serves as the core material. This combination leads to greater weight reduction without sacrificing stiffness. Additionally, faster damping of vibrations produced by these materials reduces the shock transmitted to the player's arm in tennis or racquetball games, thus providing better ball contact.

1.2.4.4 Marine applications

Glass fiber-reinforced polyester laminates have been widely used in marine applications since their introduction as commercial materials in the 1940s. In some applications, Kevlar 49 fiber is replacing glass fibers because of its higher tensile strength-to-weight and modulus-to-weight ratios. The principle of weight reduction is applied to boat hulls, decks, bulkheads, frames, masts, and spars, which translates into higher cruising speed, acceleration, maneuverability, and fuel efficiency.

1.2.4.5 What can be made using composite materials?

The range of applications is vast. A few examples are provided below:

1. Electrical and electronics: Insulation for electrical construction, supports for circuit breakers, supports for printed circuits, armors, boxes, covers, antennas, radomes, tops of television towers, cable tracks, and windmills.
2. Buildings and public works: Housing units, chimneys, concrete molds, various coverings (domes, windows, etc.), swimming pools, facade panels, profiles, partitions, doors, furniture, and bathrooms.
3. Road transport: Body components, complete bodies, wheels, shields, radiator grilles, transmission shafts, suspension springs, bottles for compressed petroleum gas, chassis, suspension arms, casings, cabins, seats, highway tankers, isothermal trucks, and trailers.
4. Rail transport: Fronts of power units, wagons, doors, seats, interior panels, and ventilation housings.
5. Marine transports: Hovercraft, rescue craft, patrol boats, trawlers, landing gear, anti-mine ships, racing boats, pleasure boats, and canoes.
6. Cable transport: Téléphérique cabins and télécabins.
7. Air transport: All composite passenger aircraft, all composite gliders, many aircraft components (radomes, leading edges, ailerons, and vertical stabilizers), helicopter blades, propellers, transmission shafts, and aircraft brake disks.
8. Space transport: Rocket boosters, reservoirs, nozzles, and shields for atmospheric reentry.
9. General mechanical applications: Gears, bearings, housings, casings, jack bodies, robot arms, fly wheels, weaving machine rods, pipes, components of drawing tables, compressed gas bottles, tubes for offshore platforms, and pneumatics for radial frames.
10. Sports and recreation: Tennis and squash rackets, fishing poles, skis, poles used for jumping, sails, surf boards, roller skates, bows and arrows, javelins, protective helmets, bicycle frames, golf clubs, and oars.

1.2.4.6 What is the world market for composites?

The global market for composites is only 10×10^9 US dollars compared to more than 450×10^9 US dollars for steel. The annual growth rate of composites is at a steady rate of 10%. Currently, composite shipments are about 3×10^9 lb annually. Figure 1.5 illustrates the relative market share of US composite shipments and shows transportation clearly leading in their use.

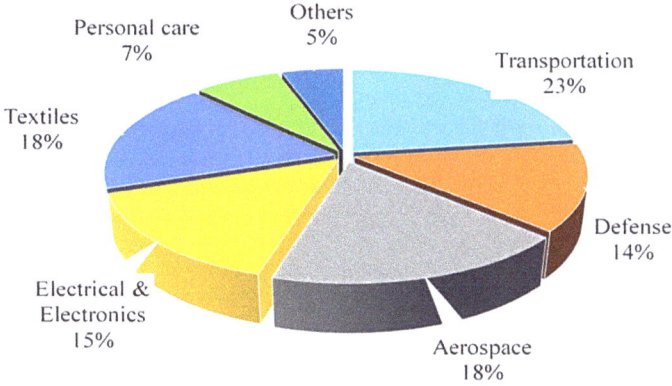

Figure 1.5: Approximate shipments of polymer-based composites.

1.2.5 Classification of composite materials

Composites can be classified either by the geometry of the reinforcement or by the type of matrix. Classification based on the geometry of the reinforcement generally includes particulate, flake, and fiber composites, as shown in Figure 1.6.

1.2.5.1 Particulate composites
Particulate composites are referred to as composites reinforced by particles having all dimensions of the same order of magnitude. They are usually isotropic because the particles are added randomly. These composites consist of particles embedded in matrices such as alloys and ceramics. The particles can be either metallic or nonmetallic. Particulate composites offer advantages such as improved strength, increased operating temperature, oxidation resistance, and many more. Typical examples of particulate composites include the use of aluminum particles in rubber, silicon carbide particles in aluminum, and gravel, sand, and cement to make concrete.

1.2.5.2 Flake composites
Composites reinforced by particles having two dimensions much larger than the third are known as flake composites. They consist of flat reinforcements within matrices. Typical flake materials include glass, mica, aluminum, and silver. Flake composites exhibit a high out-of-plane flexural modulus, higher strength, and low cost. However,

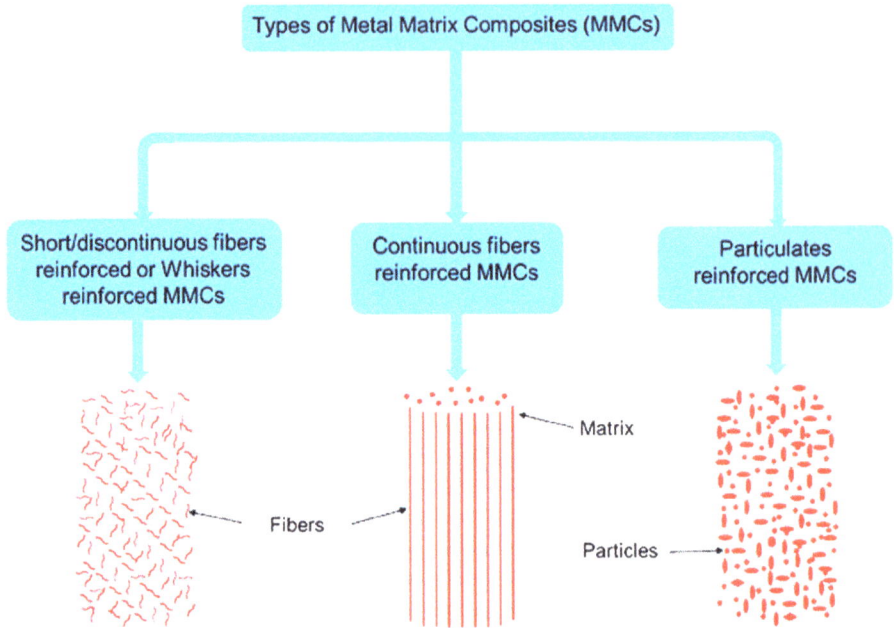

Figure 1.6: Types of composites based on reinforcement shape.

they cannot be easily oriented and are only available for use in a limited range of materials.

1.2.5.3 Fiber composites

Fiber composites consist of matrices reinforced by short (discontinuous) or long (continuous) fibers. Generally, fibers such as carbon and aramids are anisotropic. Continuous fiber composites can be made either by aligning all the fibers or by weaving a cloth and then impregnating the resulting structure with the matrix material. For discontinuous fiber composites, the materials contain short fibers, whose length is about a hundred times more than the other two dimensions. These short fibers can either be aligned or arbitrarily distributed. Short fiber-reinforced composites have great potential since the strength of the fibers increases with the decrease in fiber length and diameter. Examples of matrices include resins such as epoxy, metals such as aluminum, and ceramics such as calcium aluminosilicate.

As mentioned earlier, composites can also be classified by the type of matrix, such as polymer, metal, ceramic, and carbon. All of these types of matrices are used in composites based on the specific requirements. An important consideration in the selection of matrix material is often the service temperature. The fiber and matrix require a strong interfacial bond and should be chemically compatible to avoid undesirable reactions that can cause problems in high-temperature composites. The types of matrix composites and their service temperatures are discussed below.

1.2.5.4 Polymer matrix composites

Polymer matrix composites are the most common type of advanced composites. They consist of a polymer, such as epoxy, polyester, or urethane, reinforced by thin-diameter fibers, such as graphite, aramids, or boron. Epoxy composites are approximately five times stronger than steel on a weight-for-weight basis. This results in low cost, high strength, and simple manufacturing processes. However, polymer matrix composites have some drawbacks, including low operating temperatures, high coefficients of thermal and moisture expansion, and low elastic properties in certain directions.

1.2.5.5 Metal matrix composites

Metal matrix composites have a metal matrix, as their name implies. Metals are mainly reinforced to increase or decrease their properties to suit the needs of design. By adding fibers such as silicon carbide, the elastic stiffness and strength of metals can be increased, while the large coefficients of thermal expansion and electrical conductivity of metals can be reduced. Examples of matrices in such composites include aluminum, magnesium, and titanium, whereas typical fibers include carbon and silicon carbide. Metal matrix composites offer significant advantages over monolithic metals such as steel and aluminum. For instance, they provide higher specific strength and modulus by reinforcing low-density metals, while maintaining properties such as strength at high temperatures.

1.2.5.6 Ceramic matrix composites

Ceramic matrix composites have a ceramic matrix, such as calcium alumina silicate, and are reinforced by fibers, such as carbon or silicon carbide. This composite offers high strength, hardness, high service temperature limits for ceramics, chemical inertness, and low density. Although ceramics have low fracture toughness, their properties can be improved by reinforcing them with fibers, such as silicon carbide or carbon. Ceramic matrix composites can be applied in high-temperature areas where metal and polymer matrix composites cannot be used.

1.2.5.7 Carbon–carbon composites

Carbon–carbon composites consist of carbon fibers embedded in a carbon matrix. These composites are used in very high-temperature environments of up to 3,315 °C and are 20 times stronger and lighter than graphite fibers. Since carbon is brittle and flaw-sensitive like ceramics, reinforcing the carbon matrix allows the composite to fail gradually. The advantages are their ability to withstand high temperatures, low creep at high temperatures, low density, good tensile and compressive strengths, and high fatigue resistance. However, carbon–carbon composites face challenges such as high cost, low shear strength, and susceptibility to oxidation at high temperatures.

Chapter 2
Basic concepts and terminology

2.1 Fibers and matrix

The bond between fibers and the matrix is formed during the manufacturing phase of the composite material. This has a fundamental influence on the mechanical properties of the composite material.

As shown in Figure 2.1, most synthetic composite materials are prepared from two components: a reinforcement material called fiber and a base material called matrix material.

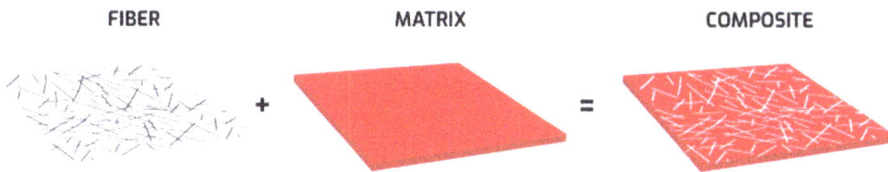

Figure 2.1: Formation of a composite material using fibers and resin.

2.1.1 Matrix materials

Although it is undoubtedly true that the high strength of composites is largely due to fiber reinforcement, the importance of the matrix material cannot be underestimated, as it provides support for the fibers and assists them in carrying the loads. It also provides stability to the composite material. The resin matrix system acts as a binding agent in a structural component in which the fibers are embedded. When too much resin is used, the part is classified as resin-rich. On the other hand, if there is too little resin, the part is called resin-starved. A resin-rich part is more susceptible to cracking due to a lack of fiber support, whereas a resin-starved part is weaker because of void areas due to the fact that the fibers are not held together and are not well supported.

2.1.2 Functions of a matrix

In a composite material, the matrix serves the following functions:
1. It holds the fibers together.
2. It protects the fibers from the environment.
3. It distributes the load evenly among the fibers so that all fibers are subjected to the same amount of strain.
4. It enhances the transverse properties of a laminate.

https://doi.org/10.1515/9783112213094-002

5. It improves the impact and fracture resistance of a component.
6. It helps to prevent the propagation of crack growth through the fibers by providing an alternative failure path along the interface between the fibers and the matrix.
7. It carries the interlaminar shear.

2.1.3 Properties of a matrix

The needs or desired properties of the matrix, which are important for a composite structure, are as follows:
1. Reduced moisture absorption
2. Low shrinkage
3. Low thermal expansion coefficient
4. Good flow characteristics ensure that it penetrates the fiber bundles completely and eliminates voids during the compacting/curing process
5. Reasonable strength, modulus, and elongation (the elongation should be greater than fiber)
6. Must be elastic to transfer the load to the fibers
7. Strength at elevated temperatures (depending on the application)
8. Low-temperature capability (depending on the application)
9. Excellent chemical resistance (depending on the application)
10. Should be easily processed into the final composite shape
11. Dimensional stability (maintains its shape)

2.1.4 Factors considered for the selection of matrix

In selecting a matrix material, the following factors should be taken into consideration:
1. The matrix must have a mechanical strength commensurate with that of the reinforcement; that is, both should be compatible. Thus, if a high-strength fiber is used as the reinforcement, there is no point in using a low-strength matrix, as it will not transmit stresses efficiently to the reinforcement.
2. The matrix must withstand the service conditions, such as temperature, humidity, exposure to ultraviolet environments, exposure to chemical atmospheres, and abrasion by dust particles.
3. The matrix must be easy to use in the selected fabrication process.
4. Smoke requirements.
5. Life expectancy.
6. The resulting composite should be cost-effective.

The fibers are saturated with a liquid resin before it cures into a solid. The solid resin is then referred to as the matrix for the fibers.

2.1.5 General types of matrix materials

In general, classification of matrix materials is shown in Figure 2.2.

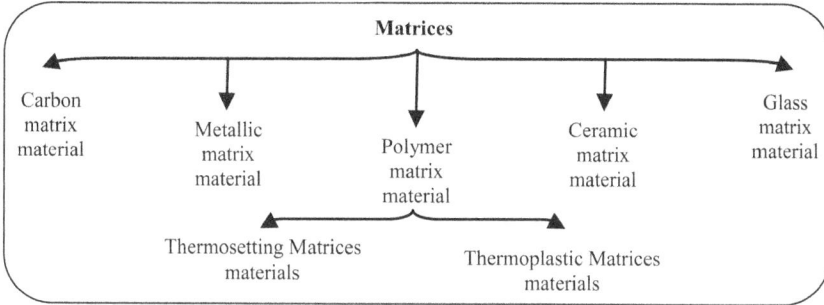

Figure 2.2: Classification of matrix materials.

2.1.5.1 Thermosetting matrices (resin)

Thermosets are the most popular fiber composite matrices, essential for research and development in structural engineering field could get truncated. Aerospace components, automobile parts, defense systems, and similar applications make extensive use of this type of fiber composites. Epoxy matrix materials, for instance, are used in printed circuit boards and related applications. Figure 2.3 shows some types of thermosets.

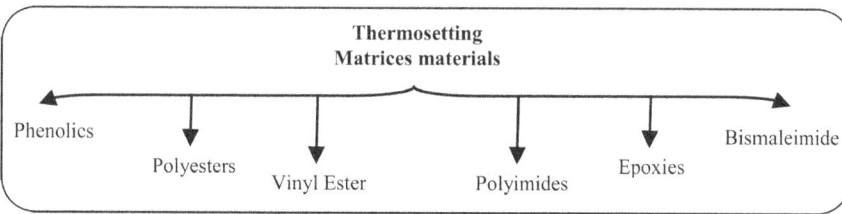

Figure 2.3: Some types of thermosets.

2.1.5.1.1 Comparison of common thermosetting resins

As can be seen, each of the resin systems has its own advantages and disadvantages. The choice of a particular system depends on the application. These considerations include mechanical strength, cost, smoke emission, temperature excursions, and other factors. Figure 2.4 illustrates a comparison of five common resins based on smoke emission, strength, service temperature, and cost.

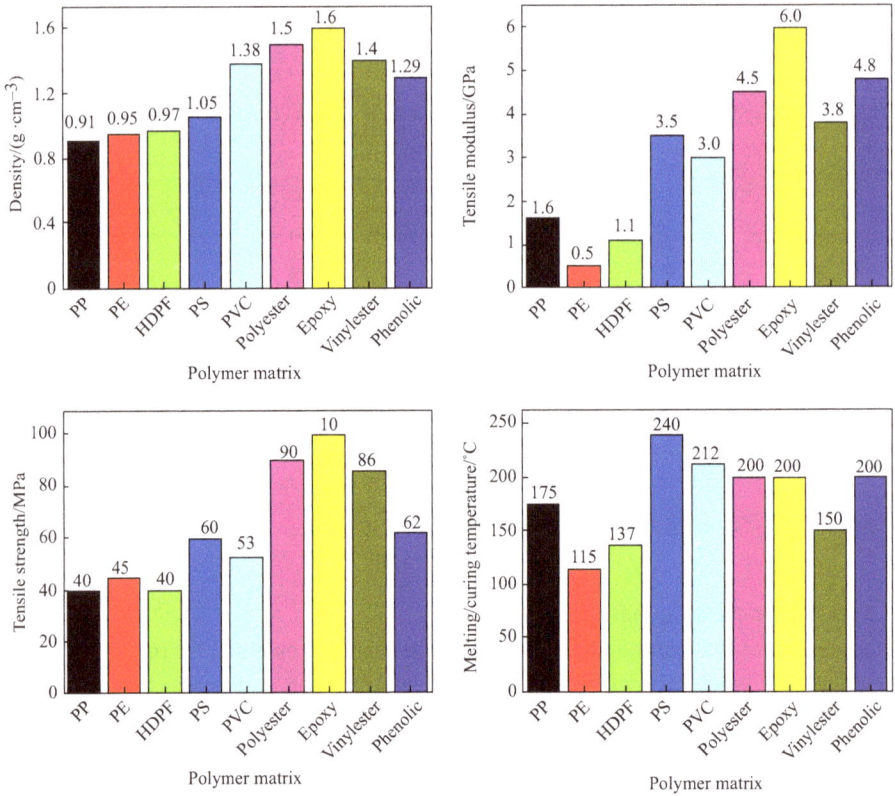

Figure 2.4: Comparison of the performance of various common matrices used in polymer matrix composites.

2.1.5.2 Thermoplastic matrices (resin)

Resins reinforced with thermoplastics now comprise an emerging group of composites. The focus of most experiments in this area is to improve the base properties of the resins and extract the greatest functional advantages from them in new applications, including efforts to replace metals in die-casting processes. In crystalline thermoplastics, the reinforcement significantly affects the morphology, promoting nucleation. Whether crystalline or amorphous, these resins have the ability to modify their creep behavior over a wide range of temperatures. However, this range includes the point at which the usage of resins is limited, and the reinforcement in such systems can increase both the failure load and creep resistance. Figure 2.5 shows some types of thermoplastics.

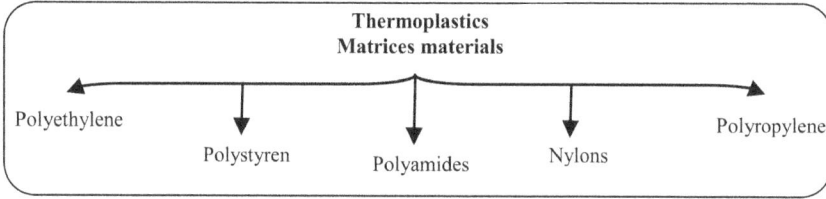

Figure 2.5: Some types of thermoplastics.

A small quantum of shrinkage and the tendency of the shape to retain its original form are also taken into account. However, reinforcements can alter this condition as well. The advantage of thermoplastic systems over thermosets is that no chemical reactions are involved, which often result in the release of gases or heat. Manufacturing is limited by the time required for heating, shaping, and cooling the structures.

Thermoplastic resins are sold as molding compounds. Fiber reinforcement is suitable for these resins. Since the fibers are randomly dispersed, the reinforcement is nearly isotropic. However, when subjected to molding processes, they can be directionally aligned.

There are a few options to increase heat resistance in thermoplastics. The addition of fillers raises heat resistance. However, all thermoplastic composites tend to lose their strength at elevated temperatures. Nevertheless, their redeeming qualities, such as rigidity, toughness, and the ability to resist creep, place thermoplastics in the category of important composite materials. They are used in automotive control panels, electronic product encasement, and so on.

Newer developments signal the broadening of the scope of applications of thermoplastics. Large sheets of reinforced thermoplastics are now available, requiring only sampling and heating to be molded into the desired shapes. This has enabled the easy fabrication of bulky components, eliminating the need for more cumbersome molding compounds.

Some of the key differences between thermosets and thermoplastics are outlined in Table 2.1.

Table 2.1: The key differences between thermosets and thermoplastics.

Thermosets	Thermoplastics
Resin cost is low.	Resin cost is slightly higher.
Thermosets exhibit moderate shrinkage.	Shrinkage of thermoplastics is low.
Interlaminar fracture toughness is low.	Interlaminar fracture toughness is high.
Thermosets exhibit good resistance to fluids and solvents.	Thermoplastics exhibit poor resistance to fluids and solvents.

Table 2.1 (continued)

Thermosets	Thermoplastics
Composite mechanical properties are good.	Composite mechanical properties are good.
Prepregability characteristics are excellent.	Prepregability characteristics are poor.
Prepreg shelf life and out-time are limited.	Prepreg shelf life and out-time are excellent.

2.1.6 Fiber materials

The fibers can be continuous or discontinuous, woven, unidirectional, bidirectional, or randomly distributed (see Figure 2.6). Unidirectional fiber-reinforced lamina exhibits highest strength and modulus in the direction of fibers, but it has very low strength and modulus in the direction transverse to the fibers. A poor bonding between the fiber and the matrix results in poor transverse properties and failures in the form of fiber pull-out, fiber breakage, and fiber buckling.

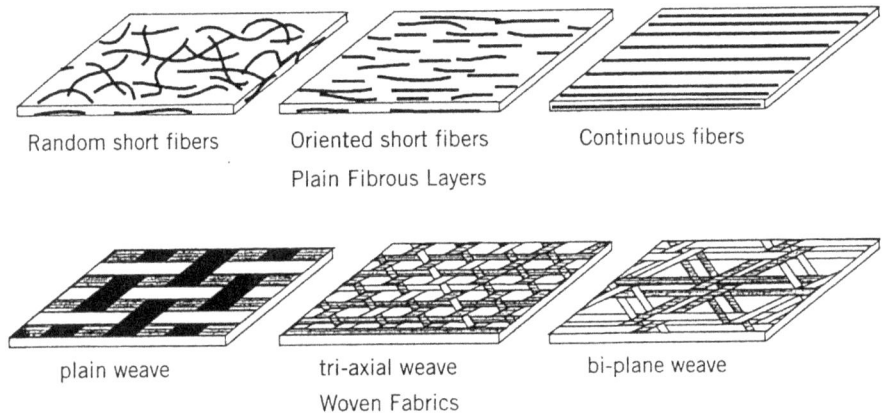

Random short fibers   Oriented short fibers   Continuous fibers

Plain Fibrous Layers

plain weave   tri-axial weave   bi-plane weave

Woven Fabrics

Figure 2.6: Various types of fiber-reinforced composite lamina.

2.1.7 Functions of a fibers

The main functions of fibers in a composite are:
1. to carry the load: In a structural composite, 70–90% of the load is carried by the fibers;
2. to provide stiffness, strength, thermal stability, and other structural properties in composites; and
3. to provide electrical conductivity or insulation, depending on the type of fiber being used.

2.1.7.1 Glass fibers

There are basically five varieties of glass fibers used in composites: E-glass, S-glass, R-glass, AR-glass, and Z-glass (zirconia-containing glass fibers). E-glass fibers are by far the most widely used glass fibers. They are used in resin matrix composites for structural and electrical applications. S-glass and R-glass fibers have superior mechanical properties compared to E-glass fibers. They are generally used in defense and aeronautical applications. AR-glass and Z-glass fibers possess good resistance to alkaline environments and are generally used as reinforcements in cement matrix composites.

2.1.7.2 Carbon fibers

Carbon fibers are prepared by the carbonization of precursor fibers in inert atmospheres at high temperatures (1,600–2,200 °C). The precursor can be an organic polymer fiber, such as rayon or polyacrylonitrile, or it can be petroleum or coal tar pitch fiber. The structure and properties of carbon fibers depend on the nature of the precursor and the conditions of carbonization. Carbon fibers derived from rayon precursors do not possess high strength. The tensile strength of these fibers can be improved by employing high-temperature stretch graphitization. This is primarily because cellulose materials yield a lower percentage of carbon upon heat treatment, resulting in higher porosity. The carbon obtained from the heat treatment of cellulose is glassy carbon, which is difficult to graphitize.

Pitch-based carbon fibers have structures very similar to that of graphite. Pitch, being a mixture of various high-molecular-weight compounds made up of fused benzene rings, yields soft carbons that graphitize readily at high temperatures. The highly graphitized structure of pitch-based carbon fibers results in a high tensile modulus for the fibers. Due to this high modulus, the fibers become extremely sensitive to the presence of defects. Thus the strength of pitch-based fibers is primarily determined by the number and criticality (size) of the flaws. The critical size of a flaw in high-modulus pitch-based fibers is estimated to be approximately 45–60 nm.

2.1.7.3 Aramid fibers

Aramid fibers are synthetic organic fibers prepared from aromatic polyamides. They are high-strength, high-modulus fibers with properties suitable for use in composite materials.

2.1.7.4 Silicon carbide

Continuous SiC monofilaments were first produced through the pyrolytic decomposition of gaseous silanes onto fine filaments of carbon or tungsten. These are thick fibers of the order of 100 μm in diameter, which continue to be of significant interest to manufacturers of metal matrix composites (MMCs) and ceramic matrix composites. An alternative production method, analogous to that for carbon described earlier, is based on the controlled thermal degradation of a polymer precursor. This process typically involves taking a precursor, such as a mixture of dimethyl dichlorosilane and diphenyl dichlorosilane, and converts it in an autoclave into a polycarbosilane from which a continuous fiber is made by melt-spinning. The fiber is then converted by pyrolysis at 1,300 °C into a fiber consisting mainly of β-SiC of about 15 μm diameter. The characteristic commercial fiber of this type is known as "Nicalon" which is marketed by the Nippon Carbon Company. It has a rough surface that promotes good fiber/matrix adhesion but is somewhat reactive toward oxygen. It is well wetted by molten metals and is reasonably stable as a reinforcement for MMCs based on aluminum and copper, although it lacks long-term thermal stability. Various attempts have been made to improve this property of the fiber, for example, by reducing its oxygen content and by adding titanium (e.g., Tyranno fiber).

2.1.7.5 Organic fibers

Bulk polymers have elastic modulus no greater than 100 MPa, but if the polymer is spun into fibers and cold-drawn to develop a high degree of molecular orientation, substantial improvements in both strength and rigidity can be achieved. In particular, polyolefin fibers with an exceptionally highly oriented, extended molecular chain structure can be produced by super-drawing in the solid state. In this process, both the crystalline and noncrystalline phases of the initially isotropic polymer are stretched and aligned, and there is an increase in crystal continuity. Such fibers have high strength, and their elastic modulus is similar to those of glass and aluminum. Apart from their excellent mechanical properties, such fibers have significant advantage over inorganic fibers that they are not brittle.

The other major development in organic fibers over the last three decades has been the production, by DuPont, of aromatic polyamide fibers, collectively known as aramids. Kevlar-49 has been the best known for many years in the composite industry. These polymers are based on oriented diamine and dibasic acid intermediates, which yield liquid crystalline solutions in amide and acid solvents. These solutions contain highly ordered, extended-chain domains that are randomly oriented in the absence of force but can be aligned by inducing shear forces in the liquid. Highly oriented fibers can therefore be produced by wet spinning these solutions, and poly (*para*-phenylene terephthalamide) fibers such as Kevlar and Twaron have strengths on the order of 2.6 GPa and modulus up to 130 GPa, depending on the degree of alignment of the polymer chains. With properties intermediate between those of carbon and glass, aramids offer an additional degree of flexibility in composite design. One important characteristic of these fibers is their extreme difficulty to cut, due to their fibrillar structure. Laser and water-jet methods are essential for trimming bare fabrics and composites containing them. Kevlar/resin composites are noted for their exceptionally high levels of toughness and resistance to impact damage. However, one disadvantage of aramids, compared to carbon fibers, is their sensitivity to moisture.

2.2 Styles of reinforcement

Many reinforcing fibers are marketed as wide, semicontinuous sheets of "prepreg" consisting of single layers of fiber tows impregnated with the required matrix resin and flattened between paper carrier sheets. These are then stacked, the orientations of each "ply" are arranged in accordance with design requirements, and hot-pressed to consolidate the composite laminates. This process is able to cope with curved surfaces, provided the degree of curvature is not too great; however, there may be a possibility of local wrinkling of the fibers when prepregs are pressed into doubly curved shapes. One way to overcome this problem is to use reinforcement in the form of a woven cloth, as textile materials can readily be "draped" over complex formers. Many fine filamentary reinforcing fibers such as glass, carbon, and SiC can be easily woven into various types of cloths and braids, the fibers being effectively placed by the weaving process in the directions required by the designer of the final composite structure. In simple designs, this may involve nothing more elaborate than an ordinary plain weave or satin weave, with fibers running in a variety of patterns but in only two directions, say 0° and 90°. However, weaving processes capable of producing cloths with fibers oriented in several directions in the plane of the cloth are readily available. Fibers of different types may also be intermingled during the weaving process to produce mixed-fiber cloths for the manufacture of some "hybrid" composites, which will be discussed later.

Most of the continuous fibers we have considered are expensive raw materials. However, the fact that the overall cost of a manufactured composite product may nevertheless be lower than that of a competing product made from cheaper, conventional materials by more costly processes makes a composite design solution an attractive alternative. Thus, although large quantities of glass fibers are supplied in chopped form for compounding with both thermoplastic and thermosetting matrix polymers, it may not initially seem economical to chop the more expensive types of reinforcement. Nevertheless, there are some advantages to use these fibers in chopped form, provided they can be arranged in the composite in a way that maximizes their intrinsically high strength and stiffness.

Chapter 3
Design concepts for composite materials/ structures

3.1 Introduction

Competent designers, engineers, and analysts can all perform their work effectively when provided with the data and design methodologies. This book is all about providing the data and design methodologies required for composite materials.

There are, however, several design considerations specific to these materials that those new to the subject need to be aware of. There is also the need for extensive interaction among the members of the design team due to the numerous influencing factors. This places increased emphasis on the role of design management in coordinating all activities to achieve an optimal solution.

It is also essential that a well-defined design process is followed in order to avoid overlapping skills and to ensure that the right skills are introduced at the most appropriate point in the design program.

The designers, therefore, need to have not only a good technical grounding in the subject but also a sound knowledge of all related activities, an understanding of how the design process works, and an appreciation for the needs of good management.

3.1.1 Design considerations

A polymeric composite material is composed of at least two materials: a fiber and a matrix. These materials are combined to leverage their individual characteristics, thereby offering additional properties that cannot be achieved independently.

They differ significantly from metals in the following ways:
1. Composites are typically orthotropic and inhomogeneous.
2. Generally, stiffness is lower than that of steel, requiring greater attention to local and overall structural stability.
3. Material properties are influenced by the manufacturing process, temperature, and the environment.
4. Most resins are combustible.
5. Material costs, particularly for high-strength and high-stiffness fibers, form a high proportion of the product cost.

Furthermore, when comparing composite materials to metals, it is found that:
1. They are lighter, resulting in excellent specific strength and stiffness values.

https://doi.org/10.1515/9783112213094-003

2. They have very good environmental resistance and do not corrode like many metals.
3. Fibers can be used strategically, leading to greater ease in optimizing weight.
4. They are readily formed into complex shapes.
5. They have low thermal conductivity.

With the above points in mind, a designer new to the subject of composite materials needs to address a number of specific areas.

3.1.2 Designing the laminate

Many structural materials typically have isotropic properties and are homogeneous; that is, to say, they are uniform in all directions.

A composite material can take a number of different forms. The material may be orthotropic, such as a unidirectionally reinforced polymer, where the strength and stiffness in the fiber direction considerably exceed those at 90° to the fiber. It may be planar-isotropic, such as a randomly chopped strand glass mat-reinforced polymer. It may approach isotropy through the use of very short fibers randomly placed in a polymer by injection molding. In all cases, however, composite materials are inhomogeneous.

It is these anisotropic properties of composite materials that are key to developing highly efficient structures. Fibers can be strategically placed to locally engineer the required strength and stiffness properties. Furthermore, by combining different fiber types – glass, aramid, carbon, and so on – the specific properties of each fiber can be exploited. For instance, the low cost of glass, the extreme toughness of aramid fiber, and the high strength and stiffness of carbon can all be used within a single laminate.

A composite material is not as ductile as metal, and its failure, when it occurs, is abrupt. The stiffness properties are generally lower than those of steel; however, the reduced weight of composite materials results in excellent specific strength and stiffness properties, enabling the creation of lightweight components and structures.

The properties of the laminate are affected by the amount of fiber in the matrix, which, in turn, is influenced by the manufacturing process.

3.1.3 Establishing property data

Obtaining reliable data for any material is essential to the success of the design function. Obtaining reliable data for composite materials is aggravated by the fragmentation of suppliers. One group of companies produces the fiber, while others produce the matrix or resins. Further companies produce yet more materials – prepregs, sandwich cores, bonding agents, and so on. Consequently, it is often quite difficult to obtain reliable data for the combination of materials that make up the finished laminate. Furthermore, the properties are influenced by the manufacturing process and the working environment.

It generally, therefore, falls to the designers to create their own database of properties by testing materials on an ongoing basis. This is an essential function within the design process, and a useful database can be gradually developed.

Reliance on manufacturers' data, where available, requires an element of caution. Often, the data are not adequately quantified, and testing standards are not made available.

3.1.4 Designing for the environment

The environmental factors of heat, cold, moisture, ultraviolet light, and aggressive materials (such as acids) can all adversely affect the performance of a composite laminate over time. The extent of the effect is a function of the fibers and resins selected and, of course, the severity of the environmental conditions.

Needless to say, composite materials perform exceptionally well in many challenging environmental situations. The marine and chemical industries have demonstrated this, where fiber-reinforced polymers are widely accepted as reliable materials. However, care is needed, particularly in a hot, wet atmosphere, where the mechanical performance of some polymeric composites can fall by as much as half from those at normal ambient conditions. This requires particular attention if long-term loading conditions are anticipated.

Adequate testing of materials in the intended environment is the only way to ensure satisfactory structural integrity.

3.1.5 Designing for joints and assemblies

Composites have the principal advantage of being chemically bonded during their manufacture. Consequently, a carefully manufactured large component does not require any joints involving mechanical fasteners or secondary bonded connections. However, it is not always possible, desirable, or practical to make a single unit. More often than not, a complete assembly is obtained from a number of components or sub-assemblies.

To this end, the designer needs to have a good understanding of the performance of mechanical joints, which are often governed by the strength of the matrix rather than the fastener itself. Consequently, long-term loads may be influenced by the creep properties of the matrix.

Alternatively, bonded joints can be used, which have the advantage of reducing load concentration through the joint. Provided care is taken to avoid stress concentrations caused by abrupt changes in section, excellent joints can be designed that are suitable for bonding and manufactured in a properly controlled environment.

3.1.6 Designing for robustness and through-life performance

Robustness is difficult to quantify, as it is a measure of the ability of a structure or component to withstand knocks, shocks, and rough handling. Composite materials can be shown to have good resilience and toughness, particularly aramid-reinforced polymers, which are also favored in ballistic missile protection. The ability of a structure to withstand through-life impacts can be enhanced by good manufacturing quality, eliminating production flaws such as voids, excessive air inclusions, and shrinkage cracks, but good design plays an equally important role. The correct selection of materials, the provision of adequate load paths, and the avoidance of stress concentrations are the most important aspects.

Composite materials generally have good fatigue resistance and they can be used to overcome through-life problems often associated with metals, particularly where the metal has been welded.

3.1.7 Designing for manufacture

A most important aspect of designing with composite materials is the interaction with the manufacturing processes, which are varied and diverse. The designer must have a thorough understanding of how these processes work and how the design will be influenced by them. For example, the quality control achievable with contact molding is significantly inferior to that available with an oven-cured fiber and resin system. Consequently, safety factors used in design must account for these differences, and the selection of design stresses may need to accommodate variations in properties that are likely to occur.

It is also essential that the design process incorporates very early involvement from those people who are truly knowledgeable about the manufacturing processes, as an incorrect choice could have disastrous consequences for the product's performance and cost.

3.1.8 Designing for cost

Expensive materials do not necessarily lead to expensive components. Lower handling costs reduce product costs, decreased weight enhances through-life performance, and increased environmental resistance prolongs lifespan, among other benefits.

Composite materials have many unique characteristics, which the designer can exploit to advantage, but cost must be carefully quantified. For any dynamic structure or component, this must include a through-life cost assessment to ensure that the best use is made of any reduced weight.

3.2 The need for design management

Design management is now a recognized and respected activity but has only recently received the attention it deserves. The driving factor has been the need to produce better and more cost-effective products. The simple objective of design management is to ensure that the product meets the set parameters and is fit for its intended purpose within an acceptable cost envelope. With the introduction of composite materials as manufacturing materials, this objective is brought sharply into focus due to the wide range of potential constituent materials – fibers, resins, sandwich cores, and so on – the varying costs of these materials, the diversity of processing techniques, and the assembly and quality control procedures. All of these factors can detract from and often confuse the design team. Design management ensures that a logical and controlled interactive procedure is followed, enabling the various disciplines to achieve the best compromise for the sake of an efficient and cost-effective product. The function and necessity of design must be understood at all levels, from the corporate to the individual. At the corporate level, the overall planning is undertaken by senior management, who defines what is needed and balances this with costs and available resources. This corporate level must also understand the importance of the design function – it is not an overhead; it is a company asset.

At the management level, project managers ensure that specific tasks are properly controlled and managed. The project objectives are defined, and the necessary disciplines are coordinated effectively. At the design level, designers use all available resources, data, and facilities to design the product within an established and controlled design methodology.

All of this may sound fairly obvious, but the design considerations mentioned above highlight the many critical factors that make management of the design so important when composite materials are involved. It is also important to understand and use a recognized design process.

3.3 The design process

It is necessary to have a framework within which design can function. The introduction of composite materials as the primary medium further underscores this necessity.

An effort to rationalize the design process for composite materials has been graphically illustrated in Figure 3.1.

The process is divided into four phases. The first phase is perhaps the most important: the brief. Without a well-defined brief, no product can be adequately designed. In fact, it could be argued that without a brief, no product should even be initiated. The effect of the brief on composite materials is important, particularly with respect to the cost. The cost of composites can vary by a factor of 100. Therefore, the brief must contain comprehensive information on the cost envelope, including through-life or operational costs. The required quantity is also important as it will directly influence the cost through factors such as tooling, cycle times, and materials used.

The cost of design is also an aspect of the first phase that requires careful attention. As previously stated, design should not be treated as an overhead but as an asset. Consequently, it should be costed and included in the overall costs. A design team working within a properly costed program is better equipped to operate within an open budget. This associated discipline is not only beneficial for the company but also helps those in the design team value their own time.

The second phase involves the preliminary design. It is during this phase that the importance of design management must be fully appreciated. There are many disciplines involved, several activities and, with composites, many different candidate materials and processes. It is extremely easy to go off in several directions when looking for an acceptable design solution, especially if those involved fail to consider the advice of their fellow professionals from other disciplines.

All designers should ask themselves during the early stages of a design program: can it be made and can it be assembled? With composites, these questions require greater attention due to the numerous processes available, each with its own advantages and disadvantages. For instance, closed-mold technology (RTM) provides a molded surface on each face, as opposed to contact molding female tooling, or prepregs. However, closed-mold technology has certain restrictions on geometry, such as draw angles, which, if not properly addressed, may lead to problems in manufacturing and assembly.

Also important in this second phase is the question of property data. One aspect of composites that is not always initially appreciated by designers accustomed to working with metals is that the material is often made at the point of manufacture, as opposed to buying in materials and then fabricated.

The material's strength and stiffness are, therefore, functions of the process, the quality control during the process operation, the skill of the operators, and, of course, the constituent materials used. The data used in the preliminary design must, therefore, be assumed or estimated theoretically. Experienced designers will have created their own databases of material properties from previous projects, enabling them to confidently select design properties. Less experienced designers must rely on published data, exercising caution when using certain data, particularly those that do not reference the source or fully quantify the properties, such as the absence of the volume fraction.

When the design concept and the manufacturing process have been finalized, the ongoing design costs are reviewed, and the brief is reassessed, the third phase begins: detailed design. During this phase, the selected materials are used with the specified manufacturing process to produce materials for testing to establish their mechanical properties, either reaffirming the estimated properties from the second phase or prompting a revision of the design based on the actual values.

The interaction of skills can perhaps be better appreciated when the second and third phases of the design process are examined. The composite industry is a fluid industry. It is constantly being introduced to new materials, processes, and end uses. No single person can be sufficiently equipped to manage all these factors simultaneously. Therefore, the interaction of skills is essential to integrate the best of these elements into the design process.

This should not, and does not, stop when the final stage is entered: that of manufacture. More so than with other materials, the design process for composite materials must continue throughout the manufacturing phase. The performance of the product is affected by the process, quality control, inspection, product maintenance, and handling. Feedback from this phase is important information to incorporate back into the design envelope.

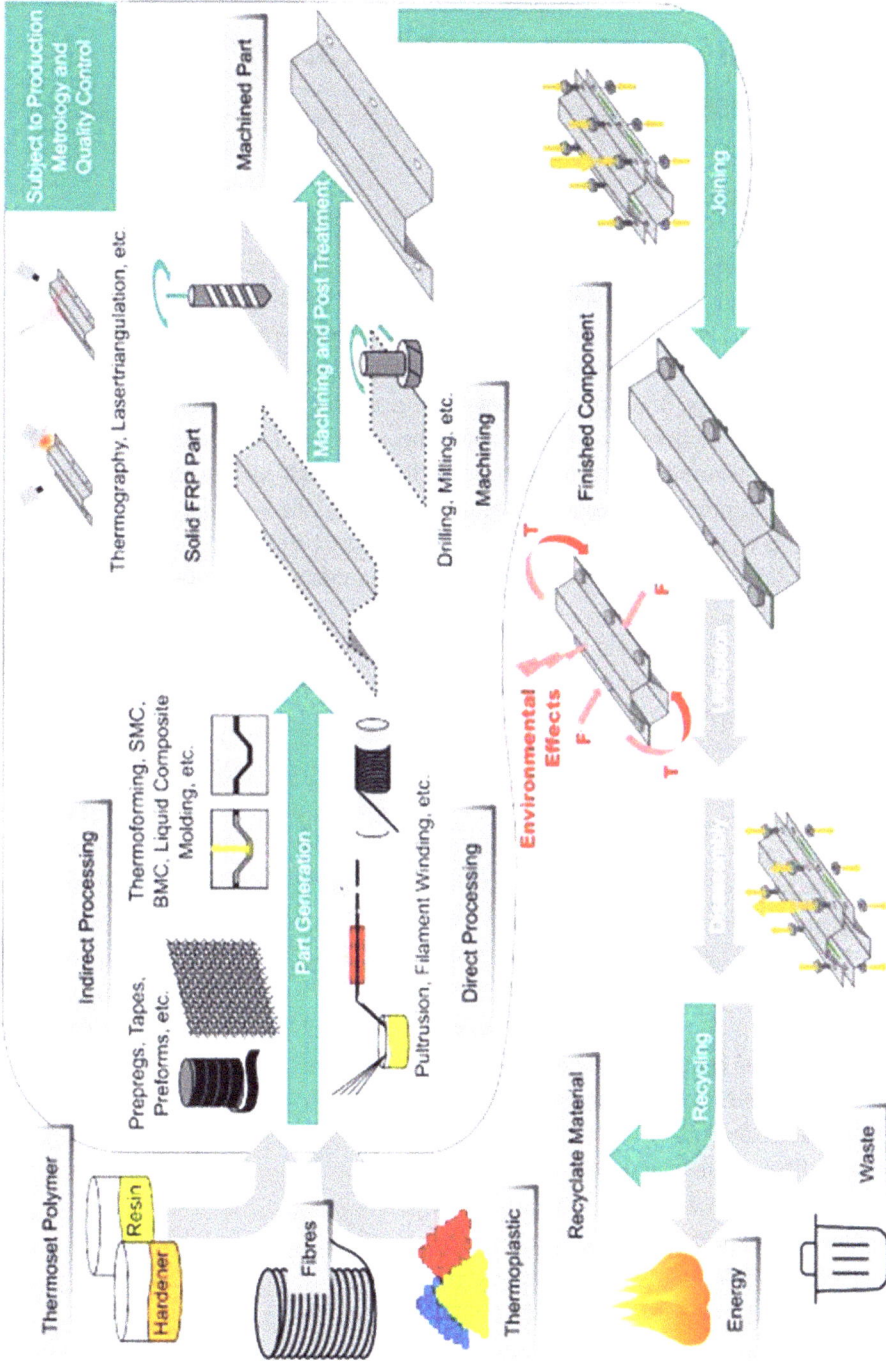

Figure 3.1: Design process for composite materials.

Chapter 4
Composite manufacturing, fabrication, and processing

4.1 Introduction

If you want to design a product using composites, there are many choices to make in the areas of resins, fibers, and core materials, each of which has its own unique set of properties. However, the end properties of a product made from these different materials are not only a function of the individual material properties. The way the materials are designed into the product and the methods used to process them play a significant role in shaping the overall end properties. The selection of a specific manufacturing process depends on several factors, including the form and complexity of the product, the tooling and processing costs, and, most importantly, the required properties of the product. This chapter describes the most commonly used manufacturing processes. The factors considered for selecting the most efficient manufacturing process are as follows:

1. User needs
2. Total production volume
3. Performance requirements
4. Economic targets
5. Size of the product
6. Labor
7. Surface complexity
8. Materials
9. Appearance
10. Tooling/assembly
11. Production rate
12. Equipment

The goals of the composite manufacturing process are to:
1. achieve consistent results by controlling
   – fiber thickness,
   – fiber volume, and
   – fiber direction;
2. minimize voids;
3. reduce internal residual stresses; and
4. process in the most cost-effective manner.

https://doi.org/10.1515/9783112213094-004

4.2 Classification of manufacturing processes

Figure 4.1 classifies the commonly used composite processing techniques in the composite industry.

4.2.1 Open mold process

Open molding offers a number of process and product advantages over other high-volume and complex application methods. These advantages include:
1. Freedom of design
2. Easy to change in design
3. Low mold and/or tooling cost
4. Possible tailored properties
5. Possible high-strength large parts
6. Possible on-site production

Disadvantages associated with the open-molding process include:
1. Low-to-medium number of parts
2. Long cycle times per molding
3. Not the cleanest application process
4. Only one surface has aesthetic appearance
5. Skill-dependent operator

4.2.1.1 Wet layup/hand layup

The simplest and oldest of the fabrication processes for fiber-reinforced polymer composites, hand layup, is a labor-intensive method suited especially for low-volume production of large components, such as boat hulls and associated parts (see Figure 4.2). A pigmented gel coat is first sprayed onto the mold to a high-quality surface finish. Once the gel coat becomes tacky, glass reinforcing mat and/or woven roving is placed on the mold, and resin is poured, brushed, or sprayed onto it. Manual rolling is then used to remove entrapped air, densify the composite, and thoroughly wet the reinforcement with the resin. Additional layers of mat or woven roving and resin are added to achieve the desired thickness. Curing is initiated by a catalyst or accelerator in the resin system, which hardens the composite without external heat.

Hand layup offers low-cost tooling, simple processing, and the potential for a wide range of part sizes. Design changes can be made easily. The parts have one finished surface and require trimming [10].

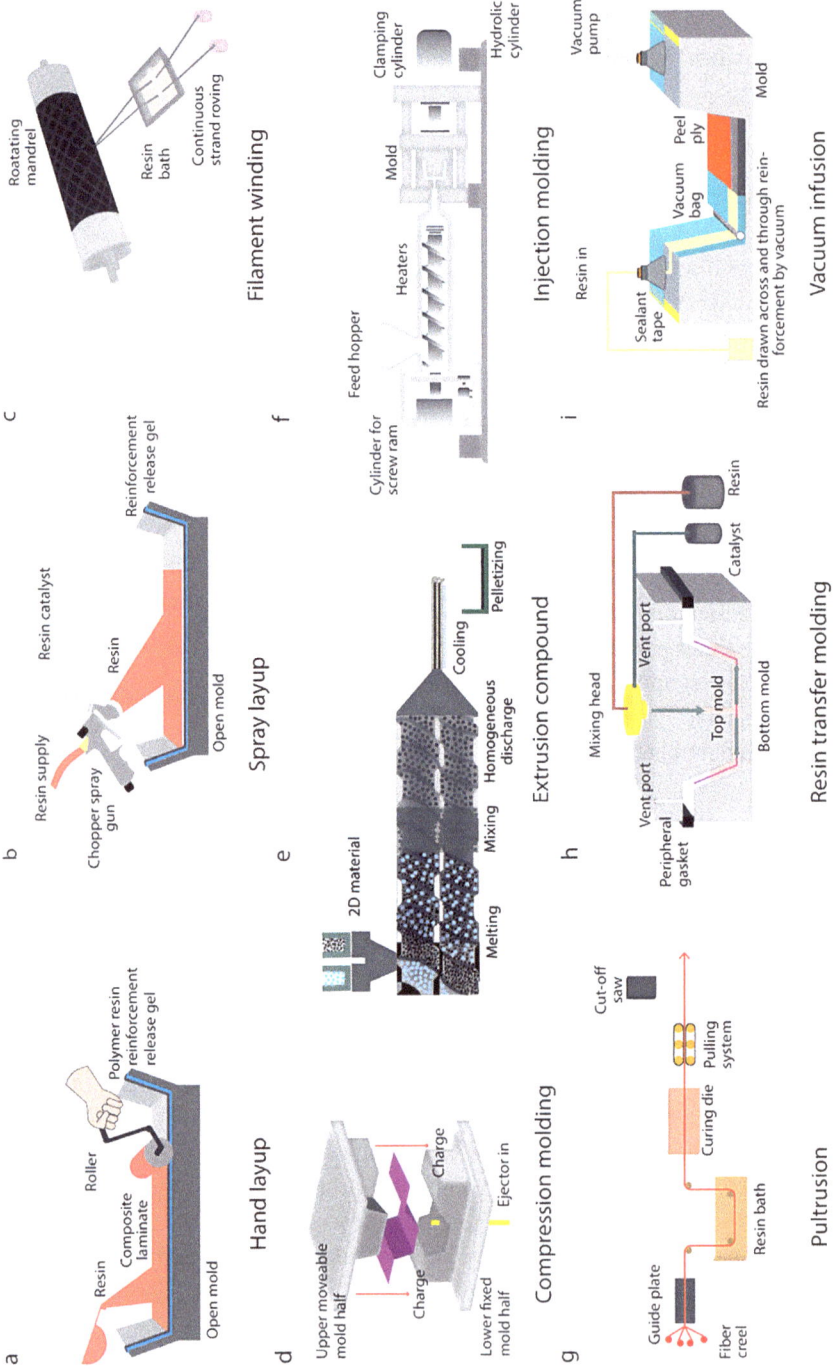

a Hand layup
Open mold
Resin
Roller
Composite laminate
Polymer resin reinforcement release gel

b Spray layup
Open mold
Resin supply
Chopper spray gun
Resin
Resin catalyst
Reinforcement release gel

c Filament winding
Roatating mandrel
Resin bath
Continuous strand roving

d Compression molding
Upper moveable mold half
Charge
Charge
Charge
Ejector in
Lower fixed mold half

e Extrusion compound
2D material
Melting
Mixing
Homogeneous discharge
Cooling
Pelletizing

f Injection molding
Feed hopper
Cylinder for screw ram
Heaters
Mold
Clamping cylinder
Hydrolic cylinder

g Pultrusion
Guide plate
Fiber creel
Resin bath
Curing die
Pulling system
Cut-off saw

h Resin transfer molding
Mixing head
Vent port
Vent port
Peripheral gasket
Top mold
Bottom mold
Catalyst
Resin

i Vacuum infusion
Resin in
Sealant tape
Vacuum bag
Peel ply
Resin drawn across and through rein-forcement by vacuum
Mold
Vacuum pump

Figure 4.1: Classification of composite processing techniques in the composite industry.

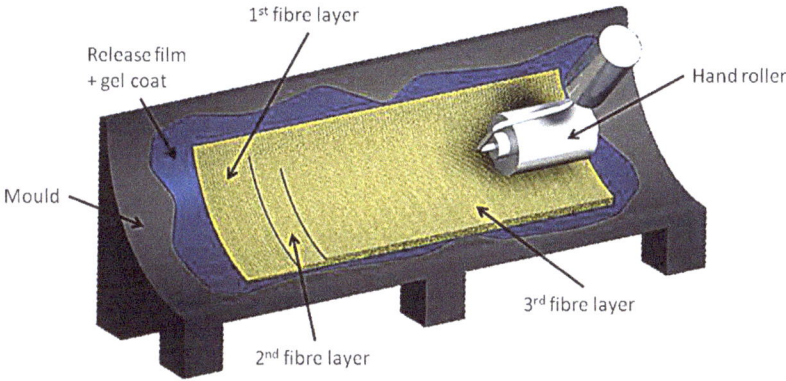

Figure 4.2: Hand layup process.

4.2.1.2 Spray layup

Similar to hand layup in simplicity, spray layup, as shown in Figure 4.3, offers greater shape complexity and faster production. It also uses a low-cost open mold (one finished part of surface), room-temperature curing resin, and is well-suited for producing large fiber-reinforced polymer parts, such as tub/shower units and vent hoods, in low-to-moderate quantities. Curing typically occurs at room temperature, but it can be accelerated by the application of moderate heat. Chopped fiber reinforcement and catalyzed resin are deposited into the mold from a combination of chopper/spray gun. As with hand layup, manual rolling removes the entrapped air and wets the fiber reinforcement. Woven roving is often added in specific areas to increase thickness or provide greater strength. Pigmented gel coats can be applied to produce a smooth, colorful surface.

Mechanical properties and off mold surfaces or contact-molding fiber-reinforced polymer components can be improved by a number of techniques. These techniques speed the cure of parts made by layup and spray-up, but they also add to processing costs.

4.2.1.3 Filament winding

Some fiber-reinforced polymer production methods involve specialized approaches to manufacturing parts that require unique properties or configurations, such as very large size, extremely high strength, highly directional fiber orientation, unusual shape, or constant cross section. In most cases, these methods are the only ones suitable for the conditions of configurations for which they were designed.

Continuous, resin-impregnated fibers or rovings are wound onto a rotating mandrel in a predetermined pattern, ensuring maximum control over fiber placement and uniformity of the structure (see Figure 4.4). In the wet method, the fiber picks up the low-viscosity resin either by passing through a trough or from a metered

Figure 4.3: Spray layup process.

application system. In the dry method, the reinforcement is impregnated with resin prior to winding.

Integral fittings and vessel closures can be wound into the structure. Once sufficient layers have been applied, the fiber-reinforced polymer composite is cured on the mandrel, and the mandrel is then removed.

Filament winding is traditionally used to produce cylindrical and spherical fiber-reinforced polymer products, such as chemical and fuel storage tanks, pipes, pressure vessels, and rocket motor cases. However, the technology has been expanded, and with computer-controlled winding machines, other shapes are now being manufactured. Today, computerized numerical control can provide up to 11 axes of motion for single and multiple spindles. Examples include helicopter tail booms, rotor blades, wind turbine blades, and aircraft cowls.

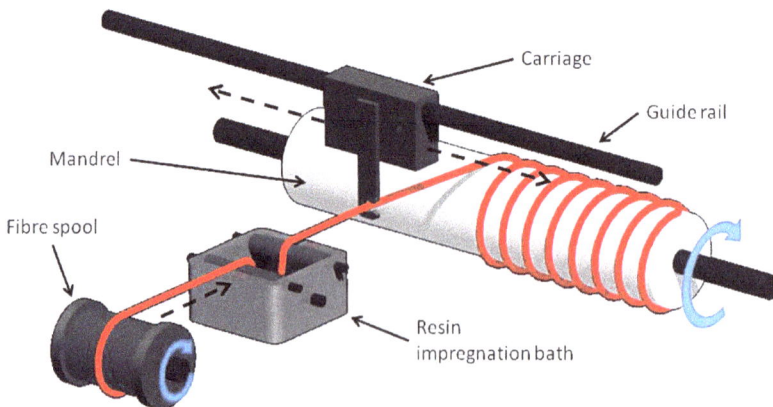

Figure 4.4: Filament winding process.

4.2.1.4 Sheet molding compound

Sheet molding compound refers to both a material and a process for producing glass fiber-reinforced polyester resin items. The material is typically composed of a filled, thermosetting resin and a chopped or continuous strand reinforcement of glass fiber. Figure 4.5 shows the molding compound in sheet form. The glass fiber is added to a resin mixture that is carried onto a plastic carrier film. After partial cure, the carrier films are removed. The sheet molding material is cut into lengths and placed onto matched metal dies under heat and pressure. Salient advantages of sheet molding compound process are as follows:

1. High-volume production
2. Excellent reproducibility of parts
3. Minimal material waste
4. Excellent design flexibility
5. Consolidation of parts

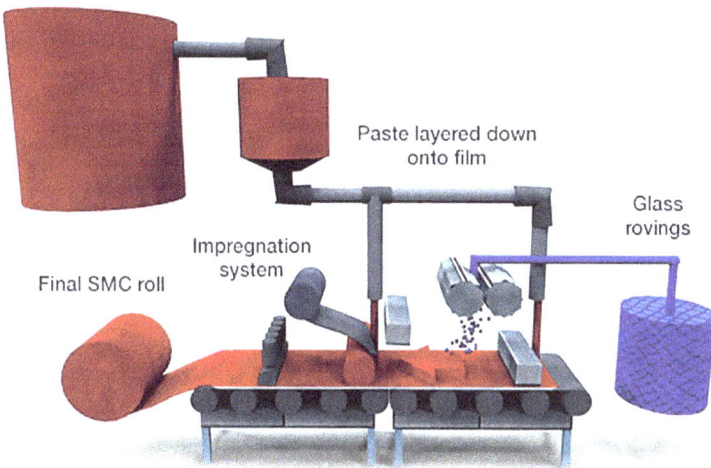

Figure 4.5: Sheet molding compound process.

4.2.1.5 Expansion tool molding

When heated to provide the molding pressure, the advantage of expansion tool molding lies in its ability to fabricate parts without the need for an autoclave. This method relies on materials with high coefficients of thermal expansion. It is designed to exploit the difference in thermal expansion between rubber and the tooling material. The female areas of the mold are made from a material with a low coefficient of thermal expansion, while the male plug is made from silicone rubber or another type of rubber tooling material having a comparatively high coefficient of thermal expansion.

When the tool is heated, the rubber male plug expands at a much greater rate than the surrounding female tool.

Pressure up to 14 MPa (2,000 psi) can be achieved at 175 °C (350 °F), which acts in all directions. The molding pressure can be regulated by adjusting the temperature, rubber composition, rubber thickness, and the ratio of rubber volume to the female mold volume.

Thermal expansion molding techniques are utilized for specialized applications involving small, complex composite structures and composite tubing with critical outer surfaces. Figure 4.6 illustrates the methods that enable the expansion of silicone rubber to provide the required pressure for compacting the composite materials.

The linear thermal expansion coefficient of most silicone rubbers falls in the range of $1-2.1 \times 10^5$. The temperature range is consistent over 23–246 °C (75–480 °F). These rubbers have a linear expansion of approximately 17 times more than that of carbon steels, which is why they are used to mold composites by thermal expansion molding techniques. Silicone rubber cools down very slowly, requiring extra time, as it cannot be removed from the composite part until it has shrunk back to its original size.

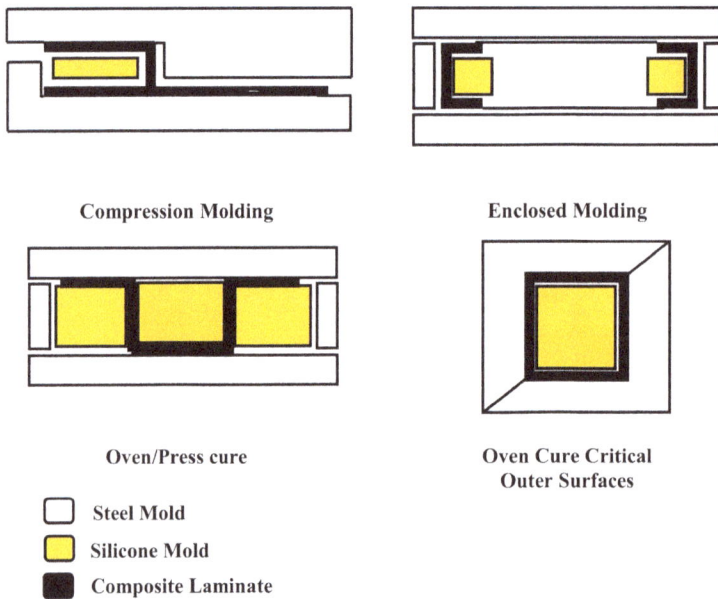

Compression Molding   Enclosed Molding

Oven/Press cure   Oven Cure Critical
                         Outer Surfaces

☐ Steel Mold
🟨 Silicone Mold
⬛ Composite Laminate

Figure 4.6: Expansion tool molding process.

4.2.1.6 Contact molding

Contact molding involves the application of molding material to an open mold, where it is allowed to cure. The process has been primarily developed for the manufacture of large fiber glass components, such as boats and automotive parts. The process requires minimal tooling and equipment costs and is, therefore, ideally suited for low-volume production. The method has two basic approaches, namely hand layup and spray layup, which differ only in the manner in which the material is applied to the mold. This method produces a high-quality surface finish on only one side of the final product.

Before the layup, the mold surface is coated with a thin layer of gel. After the layup, the part is allowed to cure at room temperature. The curing process can be accelerated by using heat lamps.

The major advantages of the method are its simplicity and low cost. However, the disadvantages of the method include the fact that only one good surface is produced, while the other side is very rough. Additionally, the method is relatively slow due to the long layup and curing time, making it unsuitable for high-volume production.

4.2.2 Closed mold process

4.2.2.1 Compression molding

For most high-volume, fiber-reinforced polymer matrix composite parts, compression molding is the primary choice. This high-pressure molding process produces high-strength, complex parts in a variety of sizes. Matched molds are mounted in a hydraulic or mechanical molding press. A measured charge of bulk or sheet molding compound, or a preform, is placed in the open mold along with a precise amount of resin. The heated mold halves are then closed, and pressure is applied. Molding time, depending on the part size and thickness, ranges from about 1–5 min. Inserts and attachments can be molded in. Compression-molded composites are characterized by good mechanical and chemical properties, superior color, and excellent surface finish. Trimming and finishing costs are minimal (see Figure 4.7).

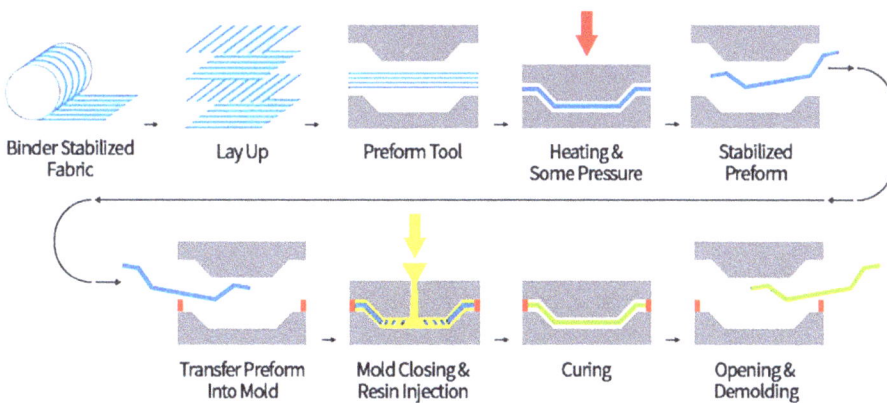

| Binder Stabilized Fabric | Lay Up | Preform Tool | Heating & Some Pressure | Stabilized Preform |

| Transfer Preform Into Mold | Mold Closing & Resin Injection | Curing | Opening & Demolding |

Figure 4.7: Compression molding process.

4.2.2.2 Vacuum bag processing

Vacuum bag processing, as shown in Figure 4.8, uses a vacuum to remove entrapped air and excess resin from a layup on either a male or female mold. A non-adhering film (usually polyvinyl alcohol or nylon) is placed over the layup and sealed at the edges.

A vacuum is drawn on the bag formed by the film, and the fiber-reinforced polymer composite is cured, either at room temperature or with heat to speed the process. Compared to the hand layup method, the vacuum method provides a higher reinforcement concentration and better adhesion between layers.

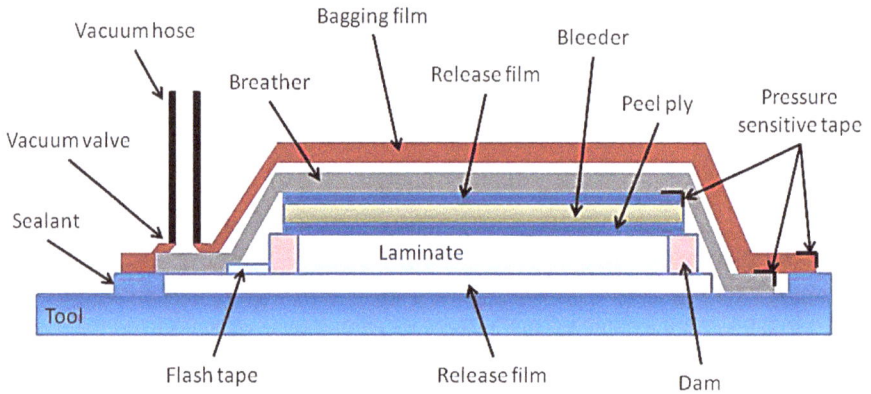

Figure 4.8: Vacuum bag process.

4.2.2.3 Pressure bag molding

Pressure bag molding is similar to the vacuum bag method, except that air pressure of 30–50 psi is applied to a rubber bag or sheet covering the laid-up assembly to force out entrapped air and excess resin. Pressurized steam may also be used to accelerate the curing process. This process is practical only with female molds (see Figure 4.9).

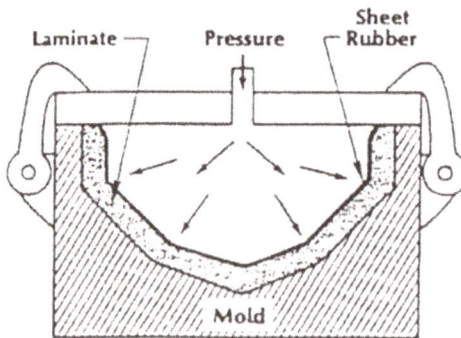

Figure 4.9: Pressure bag process.

4.2.2.4 Injection molding

Reinforced thermoset molding compounds can be injection molded (Figure 4.10) using equipment similar to that commonly used for thermoplastic resins. The principal difference lies in the temperatures maintained in various areas of the system. With thermoplastics, the injection screw and chamber are maintained at a relatively high temperature, and the die is cooled so the molded part sets up. In contrast, for a thermoset

Fiber-reinforced polymer, the screw, and chamber are cooled so that the resin does not cross-link and gel, and the die is heated so it does cross-link and cure. Injection molding offers high-speed production and low direct labor costs.

Combined with the excellent mechanical properties from long-fibered Bulk Molding Compound (BMC), the result is the capability to produce high volumes of complex parts with properties comparable to those of compression or transfer molded parts.

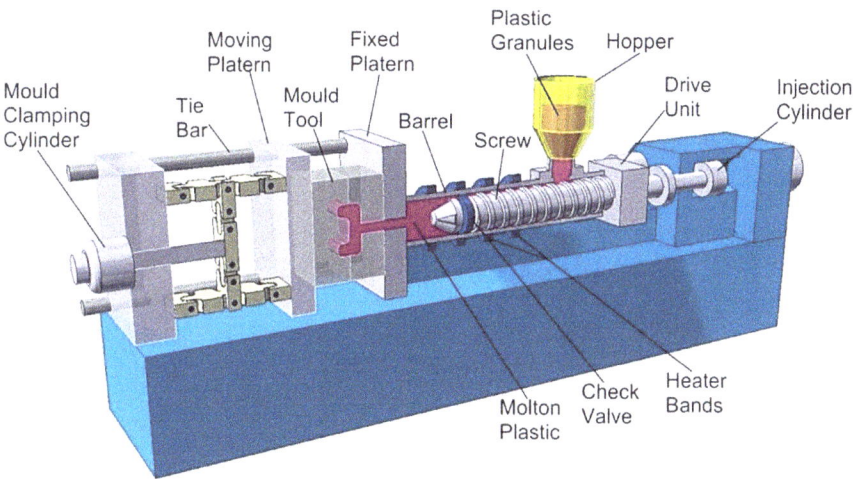

Figure 4.10: Injection molding process.

4.2.2.5 Cold press molding

Cold press molding does not use external heat to cure parts because the compound cures at room temperature, aided by self-generated exothermic heat. Cold molding is an economical press-molding method that provides two finished surfaces on the parts produced for manufacturing intermediate volumes of products using a low-pressure curing process and inexpensive molds of plaster or glass-reinforced plastic. These molds lack sharp edges, so trimming is required after molding.

4.2.2.6 Resin transfer molding

Suitable for medium-volume production of relatively large fiber-reinforced polymer components, resin transfer molding is usually considered an intermediate process between the relatively slow spray-up method and the faster compression-molding method, which requires higher tooling costs. Resin transfer molding parts, like compression-molded parts, have two finished surfaces, but molded parts require trimming. Gel coats may be used. Reinforcement mat or woven roving is placed in the bottom half of the mold, which is then closed and clamped. Catalyzed, low-viscosity resin is pumped in under pressure, displacing the air and venting it at the edges until the mold is filled. Molds for this low-pressure system are usually made from reinforced plastics.

Vacuum-assisted resin transfer molding: After the composite material is placed into the mold and the part is vacuum-bagged, a vacuum of up to 14 psi and a cure temperature of less than 350 °F are applied. The vacuum compacts the composite and helps the resin wet out the preformed composite part (see Figure 4.11).

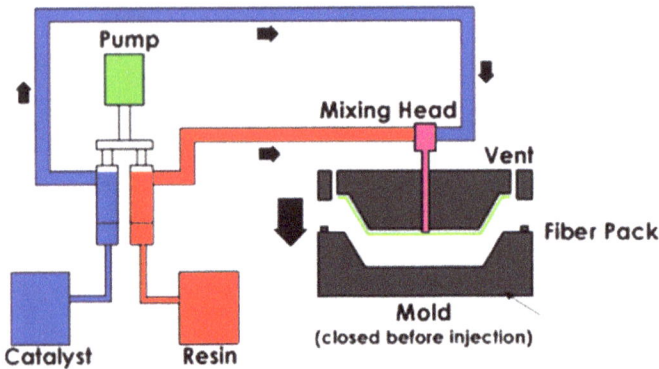

Figure 4.11: Resin transfer molding process.

4.2.2.7 Autoclave molding

Autoclave molding is a further modification of either vacuum bag or pressure bag molding. The process produces denser, void-free composites because higher heat and pressure are applied during curing. Autoclaves (Figure 4.12) are essentially heated pressure vessels (usually equipped with vacuum systems) into which bagged layups, on their molds, are taken to be cured at pressures of 50–100 psi. Autoclaves are typically used to process high-performance components based on epoxy-resin systems for aircraft and aerospace applications.

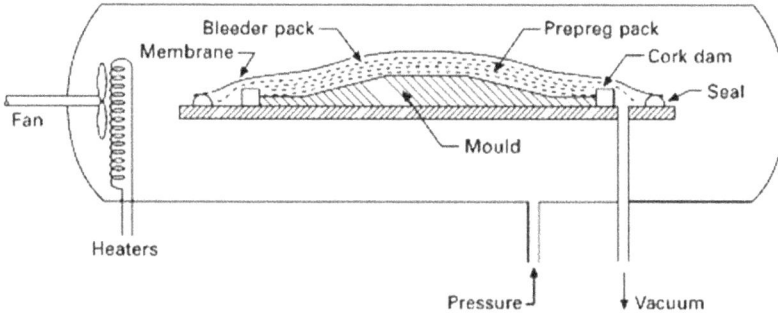

Figure 4.12: Autoclave molding process.

4.2.3 Continuous processes

4.2.3.1 Pultrusion

Constant section-reinforced fiber-reinforced polymer shapes, such as structural members (e.g., I-beams and channels), solid rods, pipes, and ladder side rails, are produced in continuous lengths by pultrusion. The reinforcement, which consists of a combination of roving, mat, cloth, and surfacing veil, is pulled through a resin bath to wet out the fibers. It is then drawn through a forming block that sets the shape of the composite and removes excess resin, followed by a heated steel die to cure the resin (see Figure 4.13). Temperature control and the time spent in the die are critical for proper curing. The finished shape is cut to lengths by a traveling cutoff saw.

Very high strength is possible in pultruded shapes because of high fiber content (up to 75%) and the orientation parallel to the length of the fiber-reinforced polymer shape. Pultrusion is easily automated, and there is no practical limit to the length of the product manufactured by this process.

4.2.3.2 Continuous laminating processes

Sheet fiber-reinforced polymer products, such as glazing panels and flat or corrugated construction panels, are manufactured using a continuous laminating process. Glass fiber chopped rovings, reinforcing mats, and fabrics are combined with resin and sandwiched between two carrier film sheets. The material then passes between steel rollers to eliminate entrapped air and establish the finished laminate thickness, followed by a heated zone to cure the resin.

Wall thickness can be precisely controlled. A wide variety of surface finishes and textures can be applied, and panel length is unlimited. Corrugations are produced using molds or rollers just prior to the curing stage.

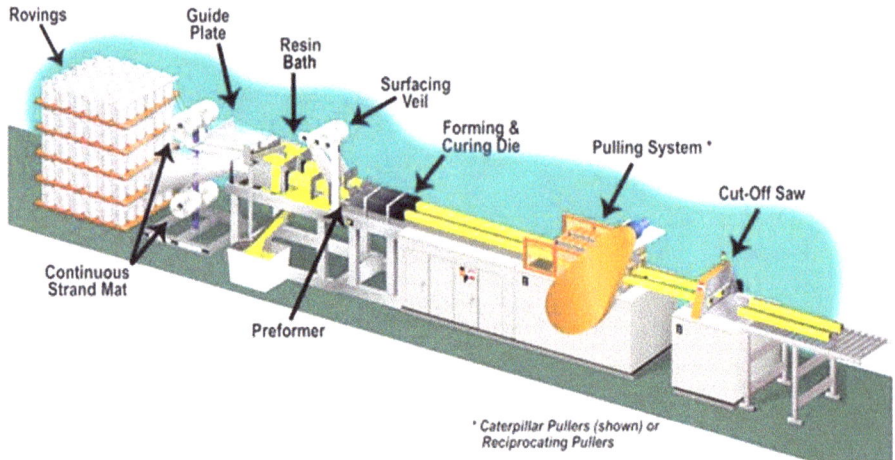

Figure 4.13: Pultrusion process.

4.2.3.3 Braiding

The braiding process involves weaving fibers into shape by repeatedly crossing them back and forth over a mandrel. The use of the braiding process in the aircraft industry is generally limited to nonstructural applications. This process has been extensively utilized for covering electrical wires and fuel lines. The primary advantage of braiding is that it provides a rapid, automated method for forming an interwoven structure. The method is a product of textile technology and usually utilizes equipment adapted from the textile industry. The braiding carriers follow a zigzag path in a large circle around the mandrel. The surface of the mandrel is tightly woven with fibers in a helical pattern.

Due to the high level of conformability and the damage resistance capabilities of braided structures, the composite industry has found structural applications for braided composites, ranging from rocket launchers to automotive parts to aircraft structures.

Two-dimensional (2D) braided structures are intertwined fibrous structures capable of forming structures with 0° and ±θ fiber orientations. Although 2D braids can be fabricated in tape form, the majority of braided structures are fabricated with a tubular geometry. Thickness is built up by over-braiding previously braided layers, similar to a ply layup process. Braiding can be performed either vertically or horizontally. A schematic of a horizontal braider is shown in Figure 4.14. Although braiding is similar to filament winding, the major difference between the two processes is that braids are interlaced structures having as many as 144 or more interlacings per braiding cycle.

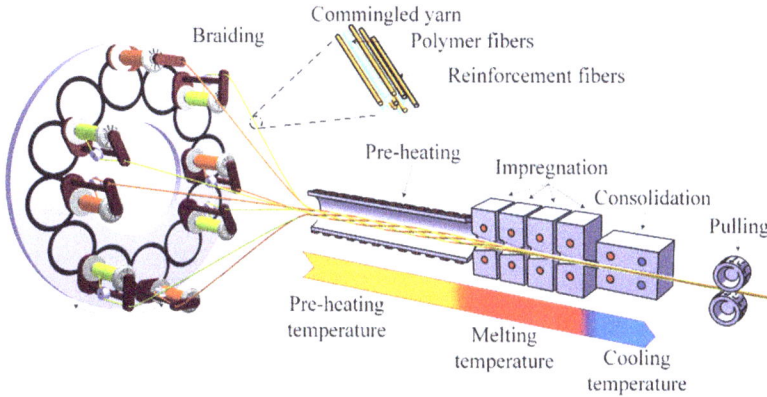

Figure 4.14: Braiding process.

4.3 Defects in manufactured polymeric composites

All practical reinforced plastic composites are likely to contain defects of various kinds arising from the manufacturing process. Indeed, composites are known for their variability in mechanical properties they exhibit unless they are produced under the most rigorously controlled conditions. The variability in materials produced by hand layup methods is more pronounced than that of composites made by mechanized processes. The specific nature and severity of the defects found in any manufactured product will also be characteristic of the manufacturing process. Additionally, any composite consisting of materials with widely differing thermal expansion coefficients that is heated during manufacture may, upon cooling, develop residual stresses sufficiently high to crack a brittle matrix. The defects that may be present in manufactured composites include:

1. Incorrect state of resin curing, especially resulting from variations in local exothermic temperatures in thick or complex sections during autoclaving
2. Incorrect overall fiber volume fraction
3. Misaligned or broken fibers
4. Nonuniform fiber distribution, resulting in matrix-rich regions
5. Gaps, overlaps, or other faults in the arrangement of plies
6. Pores or voids in matrix-rich regions
7. Debonded interlaminar regions
8. Resin cracks or transverse ply cracks result from thermal mismatch stresses
9. Disbonds in thermoplastic composites result from the failure of separated flows to re-weld during molding
10. Mechanical damage around machined holes
11. Local bond failures in adhesively bonded composite components

Chapter 5
The mechanical behavior of composite materials

5.1 Introduction

Composite materials have many mechanical behavior characteristics that are different from those of more conventional engineering materials. Some characteristics are merely modifications of conventional behavior; others are totally new and require new analytical and experimental procedures. Figure 5.1 shows the mechanical behavior of composite materials.

The most common engineering materials are classified according to their mechanical behavior.

5.1.1 Isotropic material

1. Normal stress causes extension in the direction of the stress and contraction perpendicular to it.
2. Shear stress causes only shear deformation, and this deformation is related to tensile properties.

5.1.2 Orthotropic material

1. As in isotropic materials, normal stresses cause extension only in the direction of the stress and contraction in the direction perpendicular to it.
2. Shear stress causes only shear deformation; however, shear deformation is not related to tensile behavior.

5.1.3 Anisotropic material

1. A normal stress will cause extension, contraction, and shear deformation.
2. Off-axis loading of orthotropic materials results in anisotropic behavior.
3. Samples become distorted when pulled, making it difficult to measure their properties.

https://doi.org/10.1515/9783112213094-005

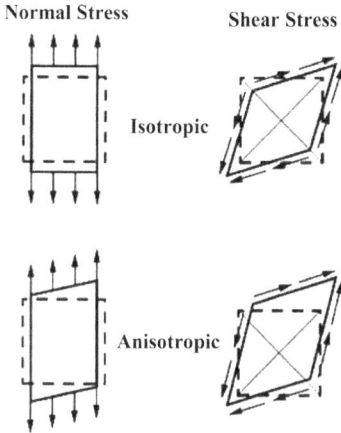

Figure 5.1: The mechanical behavior of composite materials.

5.2 Lamina and laminates

A lamina, or ply, is a typical sheet of composite material and represents a fundamental building block. A fiber-reinforced lamina consists of numerous fibers embedded in a matrix material, which can be a metal, such as aluminum, or a nonmetal, such as a thermoset or thermoplastic polymer. Often, coupling (chemical) agents and fillers are added to enhance the bonding between the fibers and the matrix material and to increase toughness. Discontinuous fiber-reinforced composites have lower strength and modulus than continuous fiber-reinforced composites.

A laminate is a collection of laminae stacked to achieve the desired stiffness and thickness. For example, unidirectional fiber-reinforced laminae can be stacked so that the fibers in each lamina are oriented in the same or different directions, as shown in Figure 5.2. The sequence of various orientations of fiber-reinforced composite layers in a laminate is termed the lamination scheme or stacking sequence. The layers are usually bonded together with the same matrix material as that used in a lamina.

Laminates made of fiber-reinforced composite materials also have disadvantages. Due to the mismatch of material properties between layers, the shear stresses generated between the layers, especially at the edges of a laminate, may cause delamination. Similarly, because of the mismatch of material properties between the matrix and the fibers, fiber debonding may occur. Additionally, during the manufacturing of laminates, material defects such as interlaminar voids, delamination, incorrect orientation, damaged fibers, and variations in thickness may be introduced. It is impossible to eliminate manufacturing defects entirely; therefore, analysis and design methodologies must account for various failure mechanisms.

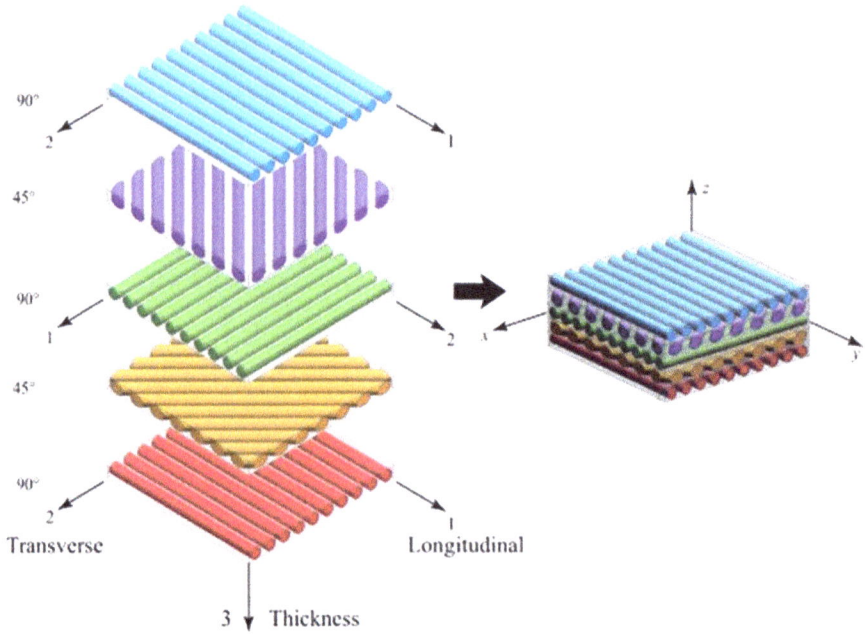

Figure 5.2: A laminate made up of laminae with different fiber orientations.

5.3 Failure modes of composite materials

In composites, the main failure modes are:
1. Microcracking of the matrix
2. Debonding (separation of fibers and matrix)
3. Delamination
4. Breaking of fibers

Figure 5.3 shows schematically different types of failure leading to damage of a laminate.

5.3.1 Microcracking of the matrix

The first form of damage in laminates is often matrix microcracking. These are intralaminar or ply cracks that traverse the thickness of the ply and run parallel to the fibers of the ply. The most common microcracking is observed in the 90° plies during axial loading in the 0° direction. These microcracks are transverse to the loading direction and are often referred to as transverse cracks. Microcracks may be observed during tensile loading, fatigue loading, temperature changes, and thermocycling. Microcracks can form in any plies but are predominantly found in plies oriented off-axis to the loading axis. The immediate

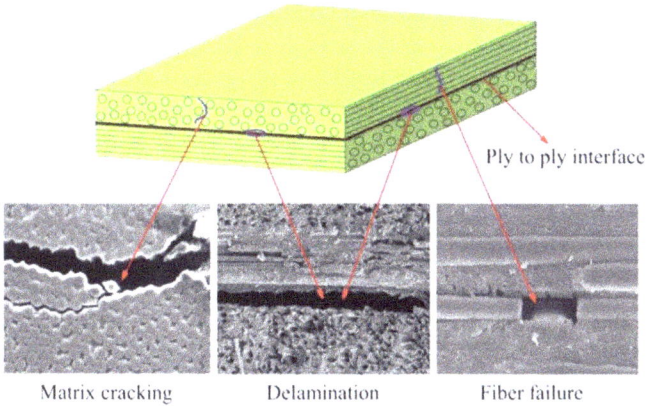

Figure 5.3: Different failure modes.

effect of microcracks is the degradation of the thermomechanical properties of the laminate, including changes in all effective moduli, Poisson's ratio, and thermal expansion coefficients. Another detrimental effect of microcracks is that they nucleate other forms of damage, such as induction of delamination, fiber breakage, or provide pathways for the entry of corrosive liquids. Such damage modes may subsequently lead to laminate failure.

The first microcrack causes very little change in the thermomechanical properties of the laminate. Continued loading, however, normally leads to additional microcracks and continued degradation in the thermomechanical properties.

5.3.2 Fiber pull-out and debonding (separation of fibers and matrix)

At some distance ahead of the crack, the fibers remain intact. In the high-stress region near the tip, they are broken, though not necessarily along the crack plane. Immediately behind the crack tip, fibers pull out of the matrix. In some composites, the stress near the crack tip may cause the fibers to debond from the matrix before they break. When brittle fibers are well-bonded to a ductile matrix, the fibers tend to snap ahead of the crack tip, leaving bridges of the matrix material that neck down and fracture in a completely ductile manner. In addition to these local failure mechanisms, when a crack reaches the interface of two laminates in a laminated composite, it can split and propagate along the interface, producing a delamination crack, as shown in Figure 5.4.

5.3.3 Delamination

Delamination is a critical failure mode in composite structures, not necessarily because it will cause the structure to break into two or more pieces, but because it can

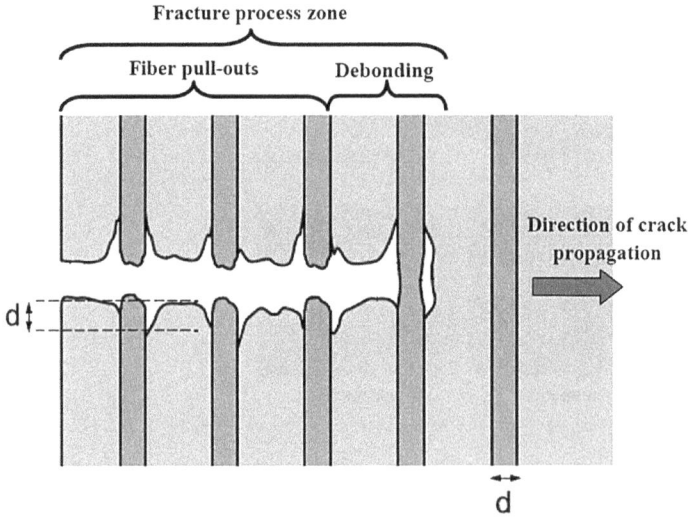

Figure 5.4: Crack tip showing localized failure events.

degrade the laminate to such a degree that it becomes useless in service. The interfacial separation caused by the delamination may lead to premature buckling of the laminate, excessive vibration, intrusion of moisture, stiffness degradation, and loss of fatigue life. The delamination though in some cases may provide stress relief and actually enhance the performance of the component.

Delamination may be introduced during processing or in service conditions. It may result from low-velocity impact, from eccentricities in the structural load path, or from discontinuities in the structures that may induce a large out-of-plane stress.

Even in the absence of such discontinuities, delamination may also result from in-plane compressive loading, which can cause global or local buckling.

The delamination may lead to a redistribution of stresses, which could eventually result in significant failure.

5.3.4 Breaking of fibers

Fiber breakage is the final stage in the failure mode of composite materials. When it occurs, the specimen is considered to have lost its structural integrity, progressing quickly to total failure. Combinations of damage modes spread within the laminates until either the tensile stress in the fibers exceeds the limiting strength of the glass, or macrocracks, which have nucleated from a combination of interlaminar and intralaminar cracks, grow catastrophically. This is often followed by the sudden failure of the structure. This type of failure is also sometimes referred to as total structural failure, as the load can no longer be supported.

5.4 Failure mechanisms of composite materials

Different types of failure modes can occur simultaneously or successively due to the anisotropic properties of the material, leading to the complete failure of the structure.

The static load failure of composites generally occurs through a combination of various mechanisms until the final failure occurs. However, in the case of laminates, different types of damage mechanisms can be observed. In general, failure in composites occurs between the matrix and the reinforcement due to defects, which are inherently present during the manufacturing process (i.e., debonding). Other common types of failure modes include matrix cracking, fiber breakage, delamination (in laminates), and buckling (under compression) [12]. Based on an extensive review of the literature, we found that the components and the size of the affected structures are the most significant parameters used to classify damages in composite materials, as presented in Figure 5.5.

In fatigue failure, the damage begins eerily, and the fatigue behavior and damage mechanisms of composites for uniaxial loaded [13–16] and multiaxial fatigue [17–19] can be understood easily. Three phases can be divided for the fatigue damage: cracking of the matrix (phase I), development of local delamination caused by cracks in the matrix (phase II), and consolidation of local delamination leading to final rupture (phase III). Also the rigidity behavior can be divided into three phases: After 10–20% of the first fatigue life (phase I), the rapid reduction triggered by cracking of the matrix, and directly next a slight quasilinear reduction (phase II). After that the damage develops rapidly, then the material will suddenly fail in the end-of-life period in fatigue (phase III) [20], as shown in Figure 5.6 [21].

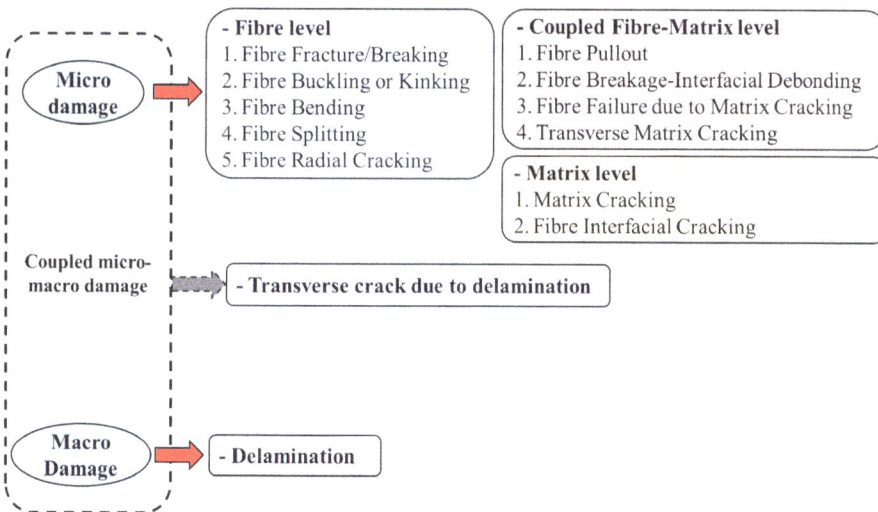

Figure 5.5: Types of damage in composite materials.

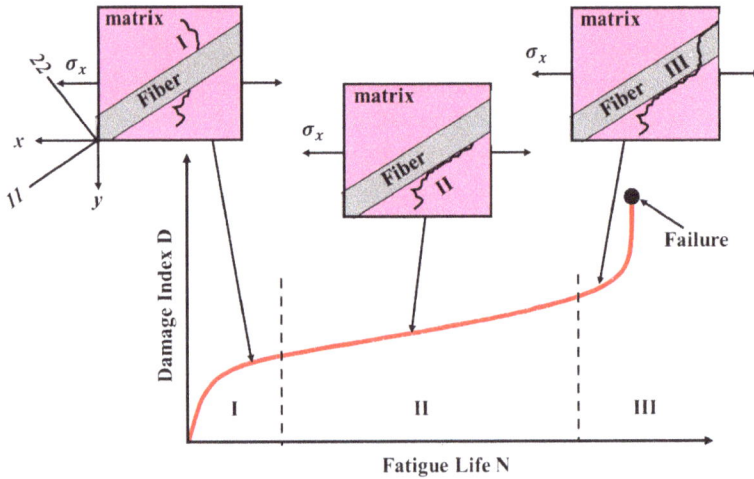

Figure 5.6: Failure mechanism in composite laminates.

Schulte et al. [22, 23] reported for the first time these three phases in reducing stiffness, which have since been observed in many types of composite materials, as well as in woven composites [24, 25].

5.5 Fatigue behavior of composite materials

The behavior of composite materials under fatigue loading is more complex due to their nonhomogeneous and anisotropic characteristics. Although this has been studied for a long time, the design of composite materials still relies on extensive fatigue testing and the use of high safety factors [26].

One of the first works on the fatigue mechanisms of unidirectional composites under axial tension–tension loading was conducted by Dharan [27, 28], who explained the roles of fibers and the matrix interface in the fatigue failure of composites. He developed a successful framework to clarify fatigue damage, known as the fatigue life diagram, as shown in Figure 5.7.

The three fatigue regions were defined individually according to the dominant mechanisms operating in these systems. Based on the characteristics of the fibers, the matrix, and the interface, the regions are positioned differently on the fatigue life diagram. As a reference construction, the three regions are placed as shown in Figure 5.8. It must be noted that Region I in the fatigue life diagram represents the quasi-static failure dispersion band, separate from the charge cycle; therefore, it does not depend solely on the matrix. However, Regions II and III are based mainly on the fatigue characteristics of the matrix but are also affected by the properties of the fibers, such as their stiffness, the fiber–matrix interface, the inelasticity of the matrix, and the fatigue load mode [29].

For example, in Region II of the fatigue life diagram, the mechanisms operating in this region demonstrate that the cyclic growth of a fiber-bridged crack will be delayed due to the presence of more rigid fibers, which leads to a longer fatigue life. In addition, stopping fatigue cracks in the matrix by stiffer fibers should be more effective, resulting in an improvement in the fatigue limit. The arrows in Figure 5.7 illustrate these trends [29]. The stiffness effect of fibers on the fatigue limit of the epoxy matrix was noted by Dharan [27]. He carried out the test data for a carbon–epoxy composite, and its fatigue life diagrams are also presented in Figure 5.7. By comparing the fatigue life diagrams of glass–epoxy and carbon–epoxy composites, it is evident that the transition from Region II to higher fatigue life occurs when moving from glass fibers to carbon fibers. Regions II and III were also found to depend on other very important factors in the fatigue life diagram. These factors must be taken into account and are organized according to their influence on the fatigue life diagram in Table 5.1, based on the authors' extensive review of the literature.

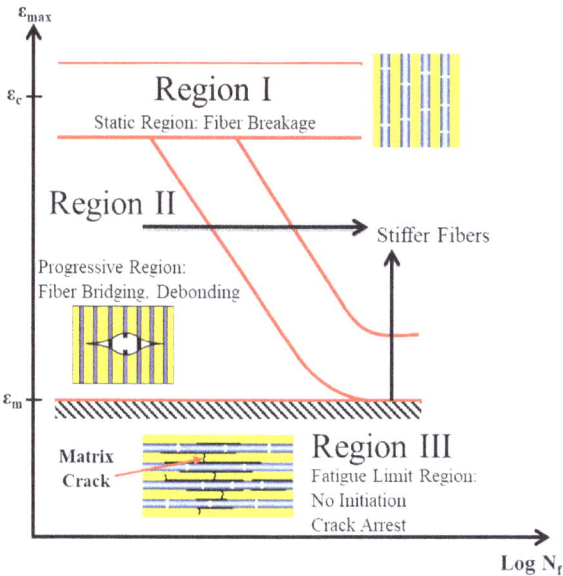

Figure 5.7: Fatigue life diagram for unidirectional composites under axial tension–tension loading.

Table 5.1: Some common factors affecting Regions II and III in the fatigue life diagram.

No.	Factors	References
1	Types of loading	[30, 31]
2	Frequency of loading	[32–34]
3	Volume fraction (V_f)	[35–37]
4	Fiber orientation	[38, 39]
5	Mean stress and stress ratio	[40, 41]

Table 5.1 (continued)

No.	Factors	References
6	Environmental factors	[42, 43]
7	Sizing and stress gradient	[44, 45]
8	Surface finish	[46, 47]
9	Stress concentration	[44, 48]

5.6 Factors affecting the fatigue behavior of composite materials

Several factors affect the fatigue behavior of composite materials. These factors must be taken into consideration and can be arranged according to their influence in this work:
1. Effect of type of loading
2. Effect of loading frequency
3. Effect of volume fraction
4. Effect of fiber orientation
5. Effect of mean stress and stress ratio
6. Effect of environmental factors
7. Effect of size and stress gradient
8. Effect of surface finish
9. Effect of stress concentration

5.7 Failure criteria of fatigue loading

Under fatigue load conditions, the material is subjected to a stress state lower than its maximum strength, so no static failure mode occurs. However, the properties of the material degrade and eventually reach the stress state level as the number of cycles increases, leading to catastrophic failure at this stage. Many researchers [49–56] have used polynomial failure criteria to predict the life of a CM subjected to fatigue load. They used the fatigue strength, based on the number of cycles, in the denominators of the failure criteria instead of the static strength of the material. This strategy may be beneficial for detecting and predicting fatigue damage. Consequently, the development of failure models to describe the mechanisms leading to failure has been studied by researchers for over 30 years. Currently, countless theories are available, which describe failure in various ways (see Table 5.2).

As listed in Table 5.2, in all the criterion formulas, all the stress components interact and contribute simultaneously to the failure of the composite systems. Therefore, the suitability of any criterion depends on the material being tested and its stress state. For all failure criteria in the formulas, the right side is the unit, while the left side consists of the local stress components divided by their corresponding strengths. The left side of any criterion is called relative damage.

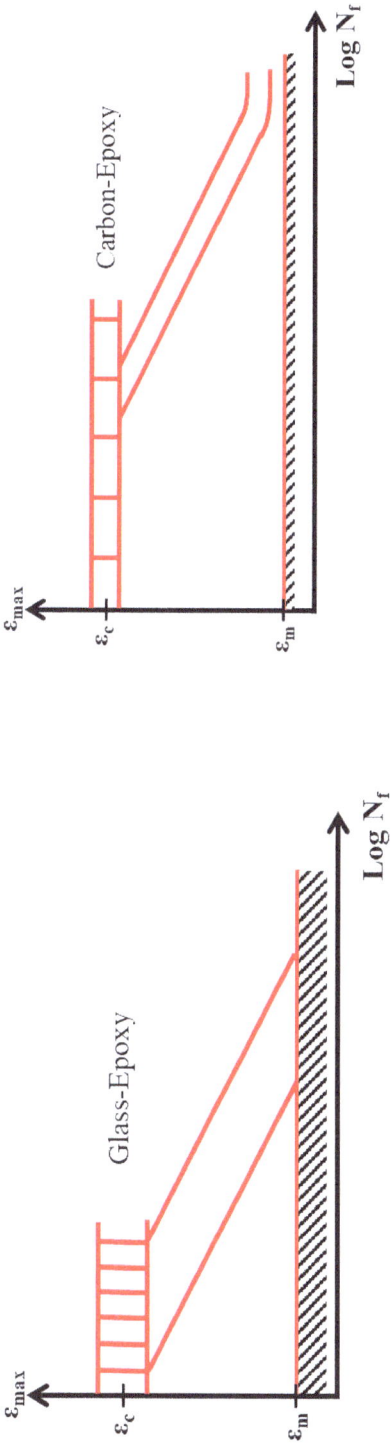

Figure 5.8: Comparison of the fatigue life diagrams for a glass–epoxy composite (left) and a carbon–epoxy composite (right).

Table 5.2: Failure criteria/theories of failure.

No.	References	Name	Mathematical formula
1	[57]	Maximum stress	$\sigma_1 = F_1$, $\sigma_2 = F_2$, $\sigma_6 = F_6$
2	[57]	Maximum strain	$\sigma_1 = F_1 + \upsilon_{12}\sigma_2$, $\sigma_2 = F_2 + \upsilon_{12}\dfrac{E_2}{E_1}\sigma_1$, $\sigma_6 = F_6$
3	[58]	Hill criterion	$\left(\dfrac{\sigma_1}{F_1}\right)^2 - \left(\dfrac{1}{F_1^2} + \dfrac{1}{F_2^2}\right)\sigma_1\sigma_2 + \left(\dfrac{\sigma_2}{F_2}\right)^2 + \left(\dfrac{\sigma_6}{F_6}\right)^2 = 1$
4	[58–60]	Tsai–Hill	$\left(\dfrac{\sigma_1}{F_1}\right)^2 - \left(\dfrac{\sigma_1\sigma_2}{F_1^2}\right) + \left(\dfrac{\sigma_2}{F_2}\right)^2 + \left(\dfrac{\sigma_6}{F_6}\right)^2 = 1$
5	[61, 62]	Norris interaction	$\left(\dfrac{\sigma_1}{F_1}\right)^2 + \left(\dfrac{\sigma_2}{F_2}\right)^2 + \left(\dfrac{\sigma_6}{F_6}\right)^2 = 1$
6	[61, 62]	Norris distortional energy	$\left(\dfrac{\sigma_1}{F_1}\right)^2 - \left(\dfrac{\sigma_1\sigma_2}{F_1F_2}\right) + \left(\dfrac{\sigma_2}{F_2}\right)^2 + \left(\dfrac{\sigma_6}{F_6}\right)^2 = 1$ or $\left(\dfrac{\sigma_1}{F_1}\right)^2 = 1$ or $\left(\dfrac{\sigma_2}{F_2}\right)^2 = 1$
7	[63–65]	Hoffman	$\left(\dfrac{\sigma_1^2 - \sigma_1\sigma_2}{F_{1t}F_{1c}}\right) + \left(\dfrac{\sigma_2^2}{F_{2t}F_{2c}}\right) + \left(\dfrac{F_{1c} - F_{1t}}{F_{1t}F_{1c}}\right)\sigma_1 + \left(\dfrac{F_{2c} - F_{2t}}{F_{2t}F_{2c}}\right)\sigma_2 + \left(\dfrac{\sigma_6}{F_6}\right)^2 = 1$
8	[66]	Modified Marin	$\left(\dfrac{\sigma_1^2 - K_2\sigma_1\sigma_2}{F_{1t}F_{1c}}\right) + \left(\dfrac{\sigma_2^2}{F_{2t}F_{2c}}\right) + \left(\dfrac{F_{1c} - F_{1t}}{F_{1t}F_{1c}}\right)\sigma_1 + \left(\dfrac{F_{2c} - F_{2t}}{F_{2t}F_{2c}}\right)\sigma_2 + \left(\dfrac{\sigma_6}{F_6}\right)^2 = 1$ where K_2 is a floating constant.
9	[67–70]	Tsai–Wu	$\left(\dfrac{1}{F_{1t}} - \dfrac{1}{F_{1c}}\right)\sigma_1 + \left(\dfrac{1}{F_{2t}} - \dfrac{1}{F_{2c}}\right)\sigma_2 + \left(\dfrac{\sigma_1^2}{F_{1t}F_{1c}}\right) + \left(\dfrac{\sigma_2^2}{F_{2t}F_{2c}}\right) + (2H_{12}\sigma_1\sigma_2) + \left(\dfrac{\sigma_6}{F_6}\right)^2 = 1$ And the following condition must be fulfilled for stability: $\dfrac{1}{F_{1t}F_{1c}F_{2t}F_{2c}} - H_{12}^2 \geq 0$
10	[71, 72]	Ashkenazi	$\left(\dfrac{\sigma_1}{F_1}\right)^2 + \left(\dfrac{\sigma_2}{F_2}\right)^2 + \left(\dfrac{\sigma_6}{F_6}\right)^2 + (2F_{12}\sigma_1\sigma_2) = 1$, $F_{12} = 0.5\left(\dfrac{4}{\sigma_x^2} - \dfrac{1}{F_1^2} - \dfrac{1}{F_2^2} - \dfrac{1}{F_6^2}\right)$ where σ_x is the global stress at 45° in tension.
11	[73, 74]	Tsai–Hahn	The same formula as Tsai–Wu, but H_{12} takes the form: $H_{12} = -0.5\sqrt{\dfrac{1}{F_{1t}F_{1c}F_{2t}F_{2c}}}$
12	[75]	Cowin	The same formula as Tsai–Wu, but H_{12} takes the form: $H_{12} = \sqrt{\dfrac{1}{F_{1t}F_{1c}F_{2t}F_{2c}} - \dfrac{1}{2F_6^2}}$
13	[76]	Fischer	$\left(\dfrac{\sigma_1}{F_1}\right)^2 - C\left(\dfrac{\sigma_1\sigma_2}{F_1^2}\right) + \left(\dfrac{\sigma_2}{F_2}\right)^2 + \left(\dfrac{\sigma_6}{F_6}\right)^2 = 1$, where $k = \dfrac{E_1(1 + \upsilon_{21}) + E_2(1 + \upsilon_{12})}{2\sqrt{E_1E_2(1 + \upsilon_{21})(1 + \upsilon_{12})}}$

Chapter 6
Structural health monitoring of composite structures

6.1 Introduction

Structural health monitoring is a new field of research and development that has evolved from smart materials and structures. Structural health monitoring has attracted considerable attention in recent years for assessment in infrastructure and aerospace vehicle applications. The goal of structural health monitoring is to develop automated systems capable of continuously monitoring, inspecting, and detecting damage to structures, thereby minimizing the need for human labor. A typical structural health monitoring system comprises three main subsystems, as shown in Figure 6.1: (i) a sensory system, (ii) a data processing system (including data acquisition, transmission, and storage), and (iii) a health assessment system (including diagnostic algorithms and information management). At a higher level, a structural health monitoring system may include a fourth subsystem having repair capabilities. The types of sensors used for damage detection are classified based on the type of data to be measured. Table 6.1 introduces various sensor types that can be used to monitor different mechanical properties of structures.

Rytter [77] described a damage detection system that has been widely accepted by both the damage detection community and the structural health monitoring community. By answering the following questions, the damage system can be described: Does the system have damage? (Existence); Where is the damage located in the structure? (Location); What type of damage is present? (Type); How severe is the damage? (Extent); and How much useful life remains? (Prognosis). In general, identifying the type and extent of damage requires prior knowledge of the structural behavior in the presence of each of the possible expected failure modes, as well as their correlation with the experimental data to be collected.

The detection of fatigue damage in composite structures can be divided into two techniques: destructive testing and nondestructive testing. Destructive testing can include the use of a variety of experimental techniques and visual on-site inspections to test samples and assess the possibility of failure, such as tensile, compression, interlaminar shear, and fracture tests. The common methods of destructive testing for analyzing damage in composite structures are sectioning, bending, and fractography.

Nondestructive testing methods have proven to be useful for the in situ assessment of composite structures, where the structural integrity of laminated composite structures can be effectively assessed. In recent years, a wide range of methods have been developed for monitoring laminated composite structures, including ultrasonic testing [78–84], X-ray radiography [85–88], thermography [89–93], acoustic emission

https://doi.org/10.1515/9783112213094-006

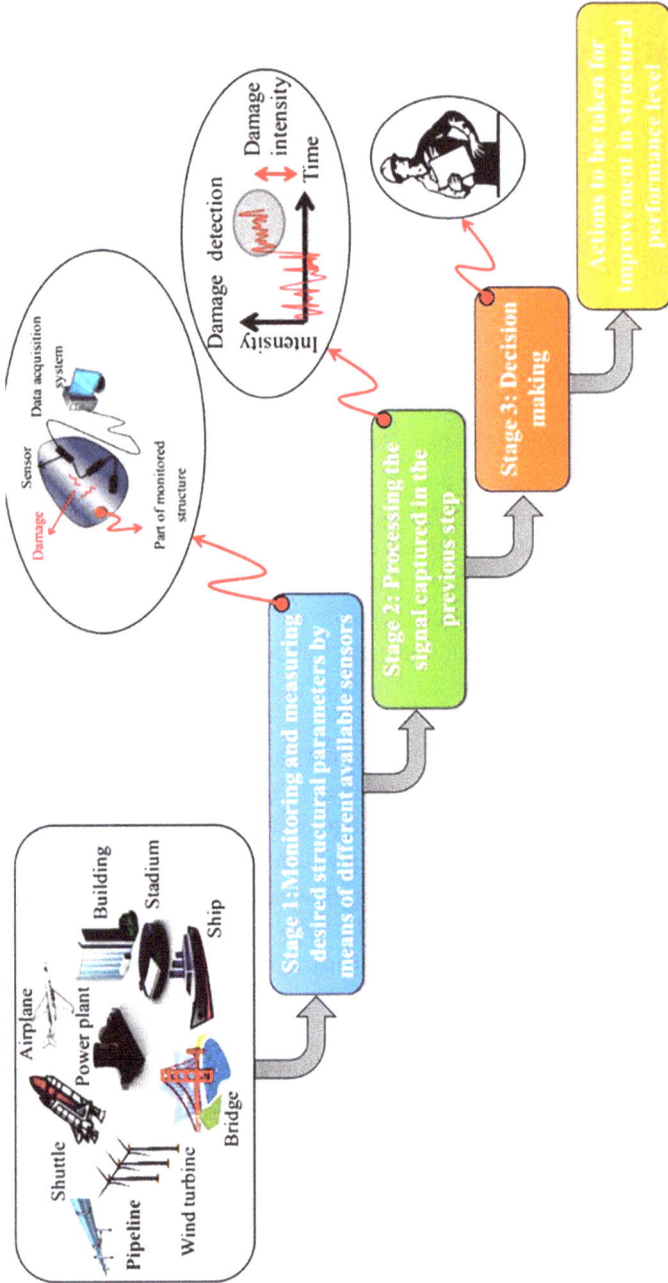

Figure 6.1: Main stages of structural health monitoring system.

[94–98], vibrography [99–107], eddy current testing [108], and optical-based nondestructive testing techniques, such as embedded optical fiber sensors [109–112], electrical potential [113–124], and Lamb wave [125–135]. Figure 6.2 lists the subcategories of
nondestructive testing types. Also Table 6.2 lists some of these methods.

6.1.1 Capability of nondestructive testing techniques for structural health monitoring

As shown in Figure 6.3, there is a wide range of nondestructive testing techniques for
detecting damage in composite structures, and the capabilities and limitations of each
method vary. Each technique has a specific area of application, although there is
some overlap depending on the assessment of the damage, the accuracy of detection,
and the ability to detect the location/size of the damage. For example, it may be necessary to combine information obtained from ultrasonic testing and X-ray radiography
to create a three-dimensional map of the complex network of location/size of delamination in a composite. However, no single method is capable of detecting all delamination parameters with high accuracy, ease, and cost-effectiveness. In addition, Figure 6.4 lists some criteria that should be considered before selecting a sensor, based
on the authors' extensive literature review.

Each nondestructive testing technique has different capabilities and limitations,
as well as specific areas of application. However, there is some overlap depending on
the size of the defect, the accuracy of detection, and the ability to detect certain types
of damage. No single method is capable of detecting all the different modes of fatigue
damage simultaneously [167]. Table 6.3 outlines the capabilities of various nondestructive testing techniques for detecting damage in composite structures. As indicated in
Table 6.3, it is crucial to use more than one nondestructive testing method to gather
as much information as possible about the different states of fatigue damage and the
residual mechanical properties of a composite. For instance, to create a 3D map of the
complex set of fatigue damage in a composite, it is necessary to combine data from
ultrasonic testing and X-ray radiography.

6.1.2 Characteristics of piezoelectric sensors for structural health monitoring

Piezoelectric sensors are a type of passive detection sensors and are commonly used
for detecting structural impacts [168]. Therefore, they can also be utilized in structural
health monitoring systems, where piezoelectric transducer sensors operate in active
mode to detect damage by generating ultrasonic guided waves. To reduce system complexity, the concept of using piezoelectric transducers for both passive and active operations has emerged. In the passive mode, piezoelectric sensors typically depend on
piezoelectric transducers to reproduce impact stress waves [169].

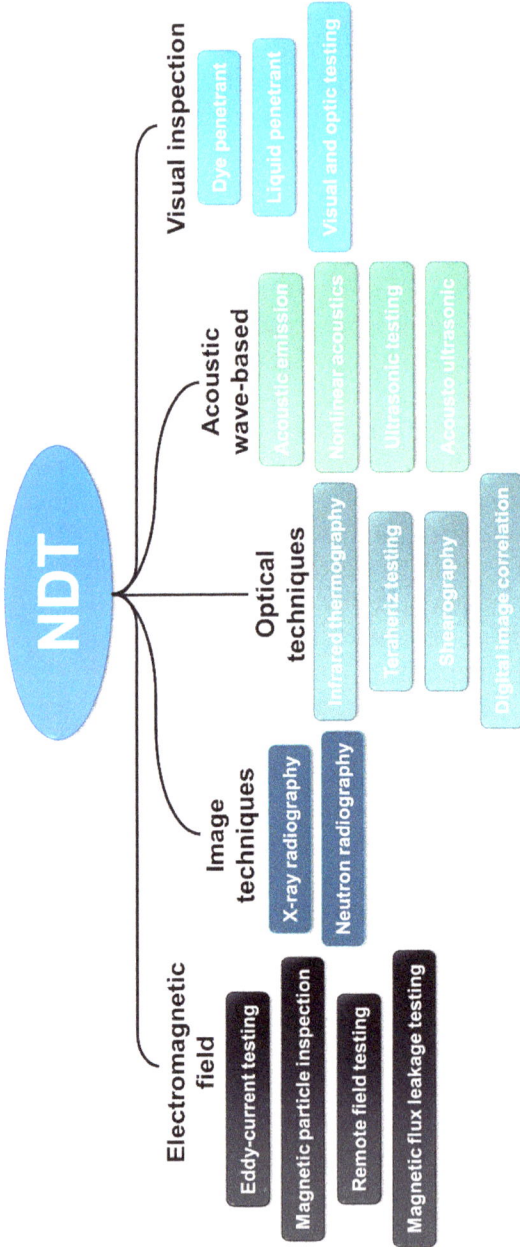

Figure 6.2: Subcategories of nondestructive testing types.

Table 6.1: Types of different sensors for damage detection of composite materials.

Measurement	Type	References
Displacement	Magnetic optical	[136]
	Ultrasonic	[137]
	Acoustic emission	[138]
	Inductive	[139]
	Capacitive	[140]
	Gyroscope	[141]
Velocity	Magnetic induction	[142]
	Optical	[143]
	Piezoelectric	[144]
Acceleration	Capacitive	[145]
	MEMS	[146]
	Piezoelectric	[147]
	Piezoresistive	[148]
Strain	Piezoresistive	[149]
	Optical	[150]
Force	Piezoresistive	[151]
	Optical	[152]
Temperature	Acoustic	[153]
	Optical	[154]
	Thermoresistive	[155]
	Thermoelectric	[156]
Pressure	Piezoresistive	[157]

Table 6.2: Some literature on recent developments in structural health monitoring of composite structures.

References	Method	Highlights	Model
[158]	Enhanced wave field imaging	The FRER index was developed to assess damage based on the amplitude subscription and residual wave components of the first Lamb waves to arrive.	A composite plate (carbon fiber-reinforced polymer, T300/3231)
[159]	Fiber Bragg grating sensors	A damage identification method for CFRP laminated plates based on strain information.	Carbon fiber-reinforced polymer laminated plates
[160]	Edge-reflected Lamb waves	The LMR of Lamb waves was proposed for structural prognosis to make it possible.	A composite plate (carbon fiber-reinforced polymer, T300)

Table 6.2 (continued)

References	Method	Highlights	Model
[161]	Frequency domain-based correlation	The CFDAC was utilized as a criterion to develop a domain-based correlation approach.	A carbon fiber-reinforced reinforced polymer laminated plate
[162]	Low-frequency guided wave	In the finite element model of different types of composite laminates, low excitation frequencies of guided waves were used for delamination detection. To obtain accurate results, two new convergence criteria were used.	A laminated composite plate
[163]	Correlation function amplitude vector (CorV)	The damage localization was evaluated accurately by combining statistical evaluation and calculating the CorV of the relative changes between intact and damaged composite laminate plates.	A composite sandwich beam
[164]	Continuous wavelet transform and mode shapes	The operational deflectional shapes or hyperspectral operational measurement systems were used for damage detection.	A composite plate
[165]	A Lamb wave-based nonlinear method	The nonlinear Lamb wave method was proposed for artificial damage detection. The damage is established between the composite laminate using a thin Teflon sheet.	A woven fiber composite laminate
[166]	Ultrasonic guided wave	To classify the delamination length, the Hilbert transform, fast Fourier transform, and wavelet transform were used to extract the effective linear and nonlinear guided wave parameters.	A composite double cantilever beam (DCB)

To classify the type of piezoelectric sensor that best suits the target application, we must first select the dimension/shape that must meet several target parameters at the system level: (1) bandwidth; (2) sensitivity/gain/signal-to-noise ratio (SNR); (3) input impedance; (4) input signal dynamics; (5) temperature range; (6) mechanical features such as stress/strain/brittleness/flexible/stretchable; (7) bonding/embedding; (8) electrical connection/wiring; and (9) cost.

There are three common and different sensor technologies using ultrasonic guided waves for damage detection. A comparison between them is shown in Figure 6.5. The most suitable type among the three types can be identified for the targeted application by comparing their electrical, mechanical, and piezoelectric characteristics.

We can define the effect of piezoelectricity as a mutual coupling between mechanical and electrical variables in a material, such as mechanical stress, strain, and electrical field

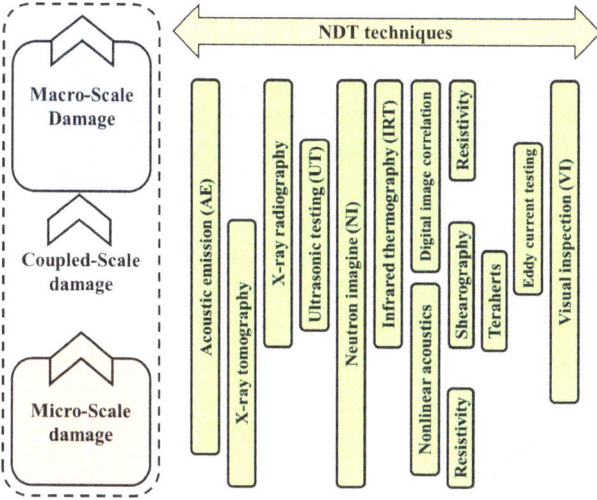

Figure 6.3: The application range of different types of nondestructive testing techniques with corresponding damage ranges.

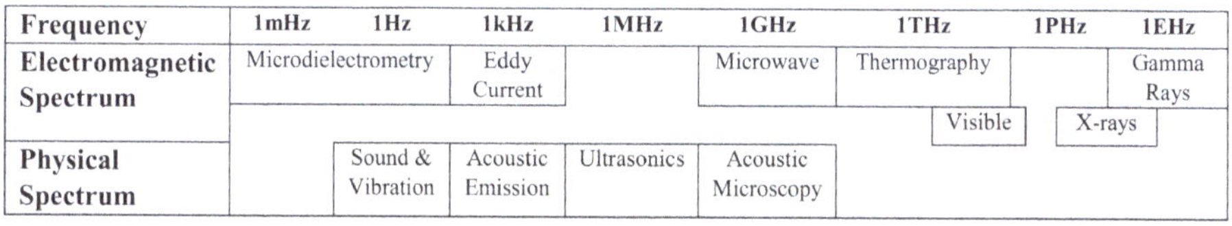

Frequency	1mHz	1Hz	1kHz	1MHz	1GHz	1THz	1PHz	1EHz
Electromagnetic Spectrum	Microdielectrometry		Eddy Current		Microwave	Thermography		Gamma Rays
						Visible	X-rays	
Physical Spectrum		Sound & Vibration	Acoustic Emission	Ultrasonics	Acoustic Microscopy			

Figure 6.4: The criteria on which the techniques of nondestructive testing and the types of sensors need to be determined.

Table 6.3: Nondestructive testing technique capabilities for fatigue damage identification in laminated composite structures.

Nondestructive testing techniques	Damage to laminates		
	Transverse matrix cracks	Delamination	Fiber fracture
Ultrasonics	Yes	Yes	No
Lamb waves	Yes	Yes	No
Acoustic emission	Difficult	Yes	Yes
X-ray radiography	Yes	Yes	Boron fibers only
Thermography	Difficult	Yes	No
Vibrography	Yes	Yes	No
Eddy currents	No	No	Boron/carbon fibers only

or charge [170]. Generally, the electrical charge is generated from a piezoelectric material by stressing it mechanically. This is called piezoelectricity [171, 172]. Thus, the 1D movement for actuation of a piezoelectric material can also be applied to deform a piezoelectric through other modes (see Figure 6.6). The piezoelectric can also be used in a reverse mechanism, where an electrical field/charge is applied to produce mechanical strain/ stress. Figure 6.7 illustrates both the direct and reverse mechanisms of the piezoelectric sensing process to monitor structures [173–175]. In this context, piezoelectric materials can function as actuators, sensors, or both [172, 173]. Figure 6.8 provides an overview of some piezoelectric-enabled detection technologies in structural health monitoring. Table 6.4 summarizes the existing studies on the types of sensors, detection modes, and working principles of piezoelectric detection techniques in structural health monitoring.

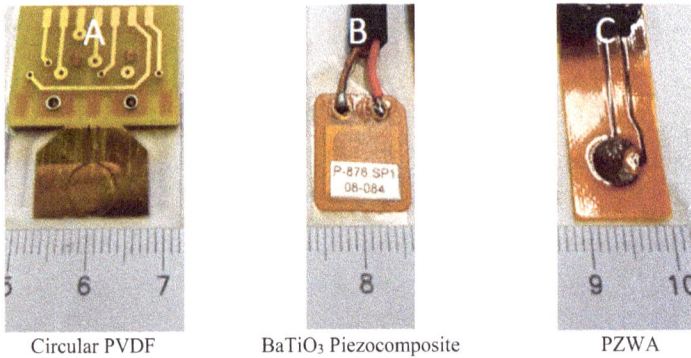

Circular PVDF   BaTiO₃ Piezocomposite   PZWA

Figure 6.5: Examples of three different types of piezoelectric sensors for structural health monitoring [169].

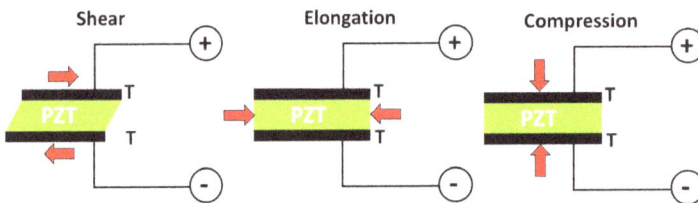

Figure 6.6: Piezoelectric deformation modes.

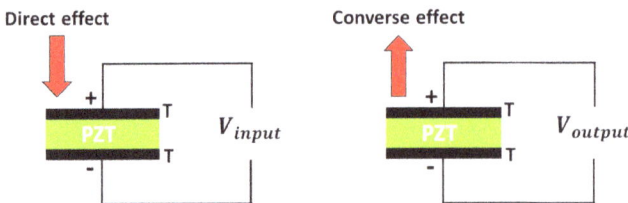

Figure 6.7: The direct and inverse response principles in the piezoelectric effect.

Table 6.4: An overview of structural health monitoring technologies based on piezoelectric materials.

Sensing mode	Sensor type	Principle	References
Electromechanical impedance	PZIT	Comparison of structures' effective resistance measuring baseline data.	[176, 194]
Guided Lamb wave	PZWA	The pulse-echo method is used for generating Lamb waves and detecting reflected Lamb waves.	[195–211]
	SMART layer technology	Exciting signals from actuators and collecting measurements from sensors, thereby detecting responses from the sensors.	[212–220]
Electrical signals	PZFG	Recording piezoelectric transducer electrical signals above predefined thresholds.	[221–225]

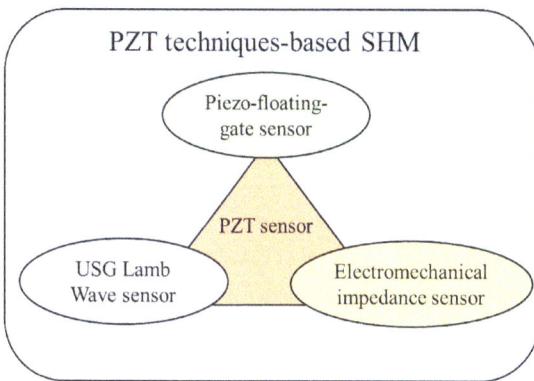

Figure 6.8: Piezoelectric transducer (PZT) technique-based structural health monitoring (SHM).

6.2 Lamb wave technique-based structural health monitoring of composite structures

Lamb waves are guided waves that propagate in plate or shell structures. There has been growing interest in using Lamb waves to identify structural damage, leading to intensive research and development in this area over the past two decades. With various engineering applications, this emerging technique serves as a tool to facilitate continuous and automated monitoring in a cost-effective manner, enhancing the integrity of engineering structures. In comparison to other nondestructive testing techniques, such as ultrasonic scanning and radiography, which have been in use for over half a century, the identification of damage using Lamb waves is still in the development phase. This approach faces a number of technical challenges in the application that need to be addressed and resolved.

The benefits of Lamb waves for nondestructive testing functions are as follows: They can travel long distances, even in materials with high attenuation, such as composites. They exhibit high sensitivity to interference along and around the path of propagation, which makes it possible to easily interrogate large areas, such as the composite wing skin of an aircraft. Lamb waves are also capable of detecting not only surface damage but also internal damage, as the entire thickness of the material can be interrogated using a variety of Lamb wave modes. Overall, damage detection methods based on Lamb waves can be used to: (1) inspect large structures without removing the coating or insulation; (2) perform 100% cross-sectional inspections of a structure over long lengths; (3) conduct structural surveys in a cost-effective manner; (4) detect multiple faults; and (5) operate with very low energy and cost [226]. As previously mentioned, using the data collected from the Lamb wave literature and following the suggestions of the relevant studies, this overview will utilize the Lamb wave technique combined with an artificial intelligence approach to detect the location, types, and severity of fatigue damage in a composite structure. This information can then be used to estimate the service life of composite structures.

6.2.1 History and literature behind the Lamb wave technique

The first discovery of Lamb waves occurred in 1917 by Horace Lamb, who defined them as elastic waves. He observed the particle movement of these waves in a plane containing the wave propagation direction, that is, perpendicular to the normal solid plates. He analyzed and described the acoustic waves of this type. The main disadvantage of Lamb waves is that their properties are quite complex. While an infinite medium only supports two wave modes moving at single velocities, thin plate-like structures such as panels, plates, and small beams with parallel free boundaries have been shown to support two infinite sets of Lamb wave modes. The velocities of these modes depend on the relationship between the wavelength and the thickness of the plate [226].

Mindlin was the first to develop a complete theory of plates in parallel with the work carried out by Schoch and Frederick between the mid-1950s and the 1960s [227]. In 1961, Worlton [228] introduced Lamb waves as a method for detecting damage. These studies collectively established the use of Lamb waves for nondestructive testing today.

Saravanos and Birman [229] conducted the first analytical and experimental application of Lamb waves in composite materials to explore the possibility of detecting delamination in composite beams. Percival and Birt [230] reached similar conclusions. Their work focused on the two fundamental Lamb wave modes. Seale and Smith [231] investigated the detection of fatigue and thermal damage in composite materials. Tang and Henneke [232] examined the sensitivity of Lamb wave propagation to fiber breakage.

Alleyne and Cawley [233] were the first researchers to discuss the interaction of Lamb waves with defects for nondestructive testing. Consequently, Saravanos et al. [234] presented a delamination detection procedure in composite materials using Lamb waves and integrated piezoelectric sensors. They used piezoelectric patches to excite the first antisymmetric Lamb wave mode.

Since then, considerable researches have been carried out on the efficiency and feasibility of Lamb waves for assessing damage in composites. Kessler et al. [235] conducted extensive studies to compare the use of Lamb waves in composites with other damage detection techniques, including Lamb wave durability and performance testing, as well as the effect of different parameters on evaluating the sensitivity of Lamb waves. Other researchers [236–242] conducted similar studies, exploring the capabilities and limitations of composite damage detection, including locating damage. These works noted that Lamb waves were more sensitive to local structural defects. Although Lamb waves have greater capabilities for detecting composite damage, a major drawback of this technique is that its application requires wave propagation continuity, a constant voltage supply, and a signal generator. Another disadvantage is the high rate of data acquisition required to achieve useful signal resolution. Finally, the mentioned studies have discovered that the Lamb wave method is significantly more sensitive and accurate for detecting and locating damage than other methods. However, the Lamb wave technique must be used in a structural health monitoring system in combination with another passive detection technique to fully benefit from its capabilities and efficiency. Examples include the frequency response function method or the spectral analysis technique, which can help save power and data storage space, as Lamb wave data may be more challenging to interpret. It has also been shown that Lamb wave data can be analyzed using wavelet transform analysis in the time–frequency domain for damage identification purposes.

Over the past decade, new research efforts have promised to develop Lamb wave-based methods that could address the aforementioned gaps and challenges associated with using Lamb waves to detect damage caused by fatigue in composite structures. For more practical and sensitive techniques to detect damage and its location, Yeum et al. [240] presented, both numerically and experimentally, a new delamination detection methodology based on Lamb waves. This methodology allows delamination detection along a single wave propagation path without requiring previous baseline data. They extracted the damage features from the circular signals of piezoelectric transducers used to excite Lamb waves.

The finite element (FE) model was also promised to simulate the Lamb wave using FE programs such as ANSYS and ABAQUS to understand the response of Lamb waves for damage identification under different degrees of fatigue damage, such as delamination and debonding. Several approaches to FE modeling have been used to simulate the propagation of Lamb waves, including modeling in the time domain, using the two-dimensional fast Fourier transform [243], wavelet transformation [244–250], short waveguide mode shapes [251], semianalytical FE models, and wavy FE models [252]. Many researchers have employed FE models to verify the results of experimental work. Shen and Giurgiutiu [135] presented a combined analytical and FE modeling approach for the accurate, efficient, and versatile simulation of 2D Lamb wave propagation and interaction with damage. In their experimental work on the propagation of Lamb waves in a specimen, Shen and Giurgiutiu used the technique of laser Doppler vibrometry. They compared large-scale multiphysical FE model simulations with Scanning Laser Doppler Vibrometry (SLDV) experiments and found that Combined Analytical Finite Element (CAFA) has excellent performance in calculating accuracy, efficiency, and versatility.

6.2.2 Lamb wave modeling and simulation

Lamb waves are ultrasonic guided waves that travel between two parallel free surfaces. As discussed, their use for damage detection has been widely explored and demonstrated. By analyzing the variations in features such as the phase/group velocity and the loss of Lamb wave amplitude in damaged versus undamaged specimens, it is possible to detect damage in materials/structures. There are two types of Lamb wave modes that propagate: symmetrical (S) and anti-symmetrical (A) modes, as shown in Figure 6.9. In the S modes, the displacements of the structure are symmetrical with respect to its central plane, while in the A modes, they are anti-symmetrical. The number of modes (S and A) is infinite in a structure [253].

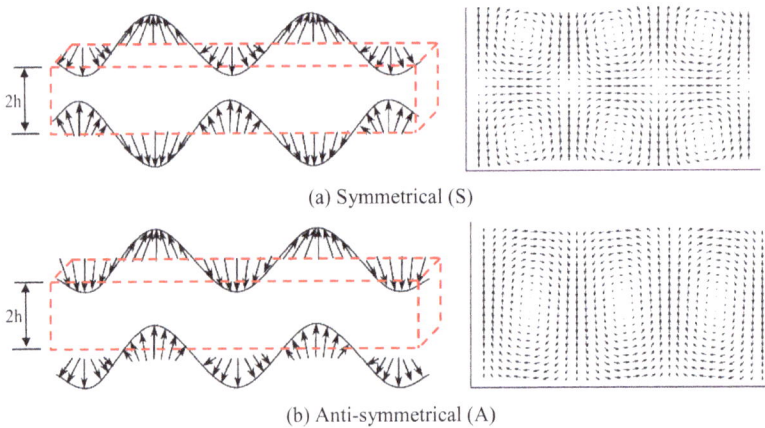

(a) Symmetrical (S)

(b) Anti-symmetrical (A)

Figure 6.9: Lamb wave modes.

Figure 6.10 illustrates the dispersion curves that describe the features of Lamb wave propagation, which are the graphs of velocities of phase/group as a function of the product of frequency thickness generated by the resolution of the equations of Lamb waves.

The dispersion curves of Lamb waves show the wave number of k as a function of the circular frequency (or the linear frequency f of the equation)). By applying the relationship to obtain the phase velocity from the wave number, the analytical expression for calculating the dispersion curves of Lamb waves in isotropic structures can be derived by solving the Rayleigh-Lamb wave equation [254], as given below:

$$\frac{\tanh(\beta h)}{\tanh(ah)} = \left(-\left[\frac{4k^2 a\beta}{\left(k^2 - \beta^2\right)^2} \right] \right)^{\pm 1} \tag{6.1}$$

where $a^2 = \left(\omega^2/c_p^2\right) - k^2$ and $\beta^2 = \left(\omega^2/c_s^2\right) - k^2$, h is the half thickness of the structure, k is wavenumber, c_p is the longitudinal wave velocity, c_s is the shear wave velocity, and ω is the wave circular frequency. The plus and minus sign in equation (6.1) is for symmetric and anti-symmetric mode, respectively [255]. The Rayleigh–Lamb waves are a type of propagates wave along a single surface. Rayleigh and Lamb waves are constrained by the elastic features of the surface (s) that guide them.

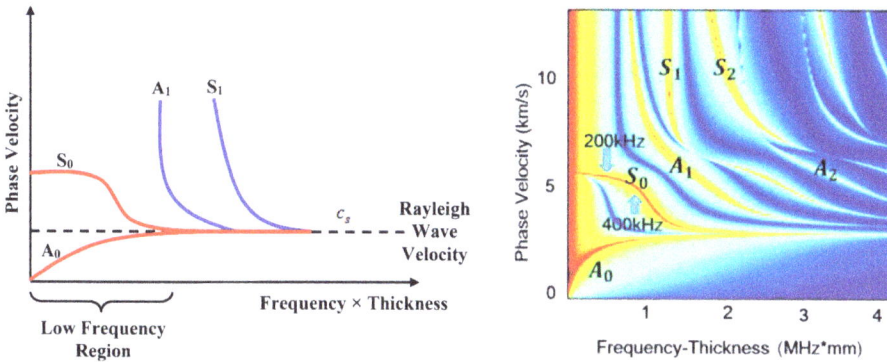

Figure 6.10: The dispersion curves describe the features of Lamb wave propagation, which are the graphs of velocities of phase/group as a function of the product of the frequency thickness generated by the resolution of the equations of the Lamb waves.

In general, the propagation of Lamb waves in nonhomogeneous composite structures is more complex than in homogeneous isotropic structures due to the properties of anisotropic materials. The characteristics of a typical anisotropic composite structure depend on the properties of the fiber and the matrix, the fiber orientation, the thickness of the lamina, and the arrangement in the thickness direction of the structure.

Two theoretical approaches to calculate the dispersion curves of composite structures are the exact solution approach and the approximate solution approach [256]. The exact solution involves applying 3D elasticity theory and solving the problem using matrix methods [257–259]. Specifically, this entails using the transfer matrix and global matrix methods. These formulations provide a matrix-based description of laminated structures in terms of stresses and displacements along the free surface, as well as the amplitudes of incoming and outgoing waves. The FE model is employed to obtain approximate solutions for the dispersion curves of waveguides with general shapes. In anisotropic composite structures, the properties of homogenized materials are utilized for each layer within the structure.

6.2.3 Lamb wave application in structural health monitoring

A passive system and an active system are two different types of Lamb waves used in structural health monitoring. In passive structural health monitoring systems, only sensors are required to detect the Lamb waves generated by damage occurring in structures. However, in active structural health monitoring systems, actuators are used to excite Lamb waves in structures, which are then picked up by sensors. The damage information is integrated into the signals obtained, as shown in Figure 6.11.

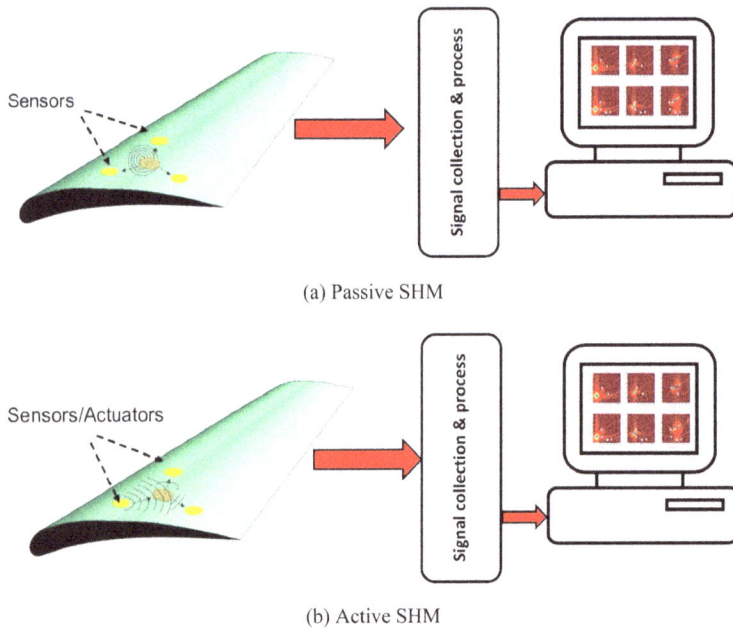

(a) Passive SHM

(b) Active SHM

Figure 6.11: The active and passive structural health monitoring systems for damage detection.

A passive structural health monitoring system mainly addresses the problem of locating damage. On the other hand, an active structural health monitoring system gives the possibility of using certain types of modes and frequency ranges, which can simplify signal processing for damage detection.

Several types of transducers have been used and developed to detect Lamb waves in composite structures, including ultrasonic transducers [260], laser transducers [261], fiber-optic transducers [262], interdigital/comb transducers [263], and speckle interferometry [264]. The most commonly used transducers with the Lamb wave technique are piezoelectric transducers, which are employed for the detection and localization of fatigue damage in composite structures. The distinguishing features of piezoelectric transducers include high force output at relatively low voltage, negligible mass and volume, simplicity of integration, a wide range of frequencies, low cost, and good response qualities at low frequencies [265, 266].

6.2.4 Optimal number and configuration of piezoelectric transducers

Optimization of piezoelectric transducers' position and number in structural health monitoring systems is an important research topic. It requires well-coordinated interdisciplinary research and should be addressed at the earliest stage of structural design. This optimization can lead to a reduction in complexity and data acquisition costs in structural health monitoring systems, while enhancing their effectiveness and quality in damage detection, as reported in the most recent review article by Ostachowicz and Soman [267]. The optimal number and position of sensors play a crucial role in accurate identification, performance maximization, and cost reduction of structural health monitoring systems. This principle applies not only to piezoelectric transducers but to all types of sensors. Table 6.5 presents the existing literature on the classification of sensor position optimization according to the structural health monitoring techniques, optimization algorithms, and application demands. However, there is a notable lack of studies addressing the optimal number and position of sensors in structural health monitoring systems.

Optimal piezoelectric transducer placement in the Lamb wave method for damage detection is both desirable and challenging. The goal is to reduce the number of sensors to keep costs and the complexity of data analysis low. However, this can only be achieved if the sensors are effectively placed to detect and classify damage at all positions on the structure. The genetic algorithm is the most suitable method for optimizing these problem. This technique is well-known to work based on biological natural selection, whereby a population of potential solutions evolves over many generations to converge on an optimal "fit" solution. The full definition of the genetic algorithm is provided by Katoch et al. [307]. Briefly, it offers a solution for damage detection.

In this case, the piezoelectric transducer locations, represented as a string of integer numbers, are encoded analogous to the genetic code on a DNA string. Accordingly, the piezoelectric transducer output of guided waves propagates over different ranges of frequencies, which is analogous to the behavior of a breeding population consisting of a number of individuals. Each characterized by its DNA and each individual is determined according to some fitness function. Here, it is preferable to prioritize individuals with high fitness during the breeding process, so that useful genes tend to propagate through generations while detrimental genes disappear similar to the range of frequencies generated by the piezoelectric transducer.

In this way, a large area of research can be explored readily, quickly, and with high accuracy in finding a global optimum. Usually a genetic algorithm code is written using MATLAB and by using FE package such as ANSYS to calculate the voltage frequency response from both intact structures and various damaged structures. Figure 6.12 illustrates the basic framework of the genetic algorithm. For more details on applying genetic algorithms for optimal piezoelectric transducer placement, refer to the work by Daraji et al. [297].

Table 6.5: The classification of optimization of sensor position.

Classification items	Method	References
Structural health monitoring techniques	Vibration monitoring	[268, 269]
	Strain monitoring	[270–273]
	Wave propagation-based monitoring	[274–279]
Optimization algorithms	Biology-based algorithms	[280–286]
	Physics-based algorithms	[287–289]
	Geography-based algorithms	[280]
	Sequential placement	[290]
	Genetic algorithm	[291–296]
Application demands	Wireless sensors	[286]
	Sensor types	[273]
	Piezoelectric	[297–303]
	Multi-type sensor networks	[304–306]

6.2.5 Lamb wave-based damage identification principle

The most suitable method for damage detection based on Lamb waves must be able to take into account noise, structural vibrations, and the overlapping of several modes. The methods used for detection can be divided, based on the nature of the signal processing approach, into three main categories: time domain analysis [308], frequency domain analysis [309], and time–frequency domain analysis [310].

In most cases of detecting fatigue damage in composite structures using the Lamb wave technique, the time domain-based method for processing the wave signal is the most appropriate, whereas the damage is evaluated using the histories of the time of the input and we can detect damage proceedings more easily, both locally and globally, as shown in Figure 6.13.

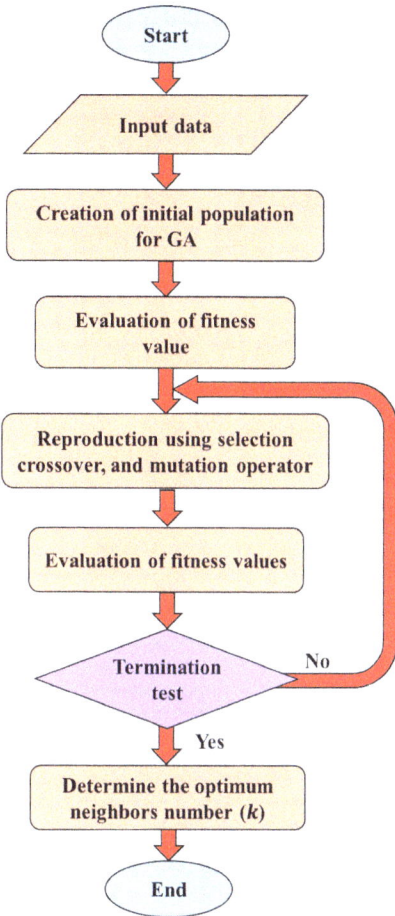

Figure 6.12: The basic framework of a genetic algorithm.

There are two piezoelectric transducers. One of them generates ultrasonic guided waves, which propagate through the structure, pass through the damaged area, interact with it, and collect information about the damage. This information is then picked up by another piezoelectric transducer. For structural discontinuities, where damage crosses or approaches one or more sensor paths, the wave signals propagating along the affected path are modulated, and damage information can be obtained from the detected signals along the sensor path. The signal propagation does not depend on the damage size, where the ultrasonic guided waves can propagate with minimum damage, same sensitivity, and large damage size, because when the signal encounters any discontinuity in structural geometry, the wave will be scattered, reflected, and mode-converted [311].

The important parameters of this technique are the time of flight (ToF) of wave and the time window of the guided wave signal, which are the primary focus of research in damage detection where ToF of the wave refers to the signal's flight time from an actuator to a sensor. It can be used to compute the distance and the velocity of wave propagation.

There is also the time window, which is very important to calculate. It is the main component of the received signal, characterized by relatively large amplitude, as well as containing more information and energy. The time window refers to the period between T_{start} and T_{end}, which are given in Equations (6.2) and (6.3), respectively:

$$T_{start} = T_1 + \text{ToF} - \frac{1}{2}T_0 \tag{6.2}$$

$$T_{end} = T_1 + \text{ToF} + \frac{1}{2}T_0 \tag{6.3}$$

where T_1 represents the wave excitation time of an actuator, which is the middle time point of T_0 time window, time of wave flight is the time of flight from an actuator to a receiver, and T_0 is the period of the excitation wave envelope. The time window is illustrated in Figure 6.14.

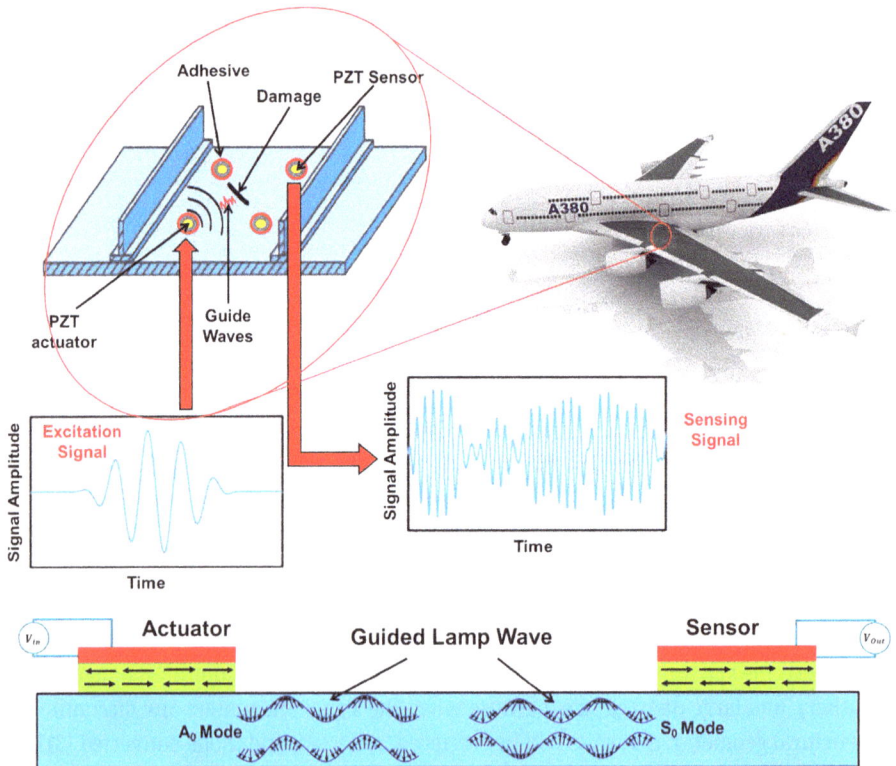

Figure 6.13: Damage identification principle based on Lamb waves.

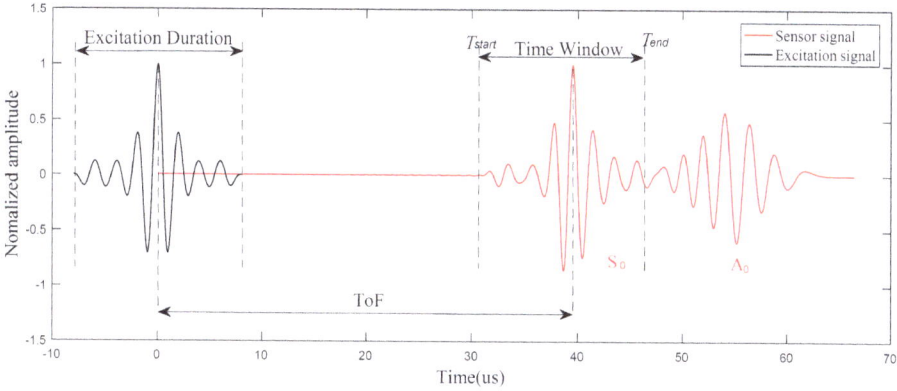

Figure 6.14: Schematic illustration of the time window calculation.

The damage detection strategy in the analysis of the time domain of wave signals depends on cross-correlations with the measurement of baseline data used to represent the signal (no damage conditions). The function of these baselines is to provide a comparison with the boundary effects of wave signals (damage conditions), and it is difficult to maintain and change the data of these baselines with environmental conditions, as shown in Figure 6.15.

Two detection methods based on the time domain approach include the direct time method and the time inversion method. In the direct time domain method, the history of Lamb waves propagating through the structure is usually recorded from the time-series signals to provide direct information about the waves (see Figure 6.15), as shown in Figure 6.15. When the piezoelectric excitation signal propagates through the damaged area, it interacts with the damaged edges, and changes the wave amplitude and propagation path and wave packet arrival time increase of Δt. Subsequently, the wave frequency of the guided wave changes due to the expansion of the damaged area and change in structural thickness, so that the guided wave velocity and arrival time are also affected. The factors influencing the response signals include input signal power, signal attenuation, the geometry of the monitored structure, the sensor array, and other related parameters.

Due to factors such as the existence of various wave modes, propagation speed, attenuation, distance dispersion, scattering, or damage from structural boundaries. Firstly, the time reversal method has been used as a method to compensate for Lamb wave dispersion; then recently it has been used to damage detection. Compared with traditional methods, the advantage of using the time reversal method is that it does not require a baseline database [312]. This is particularly beneficial as changes in environmental and operating conditions can be accounted for, helping to minimize false positive damage alerts. Currently, the time reversal method has been successfully implemented to detect the presence and location of damage, and it is also capable of determining the extent and type of damage.

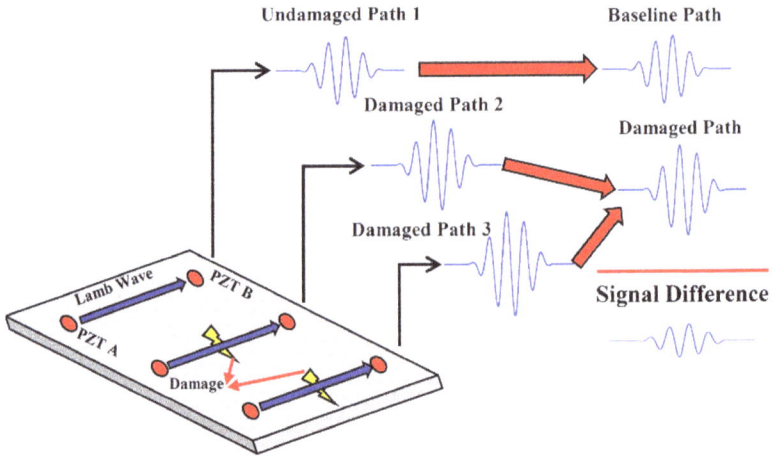

(a) Baseline data SHM Method using lamp waves.

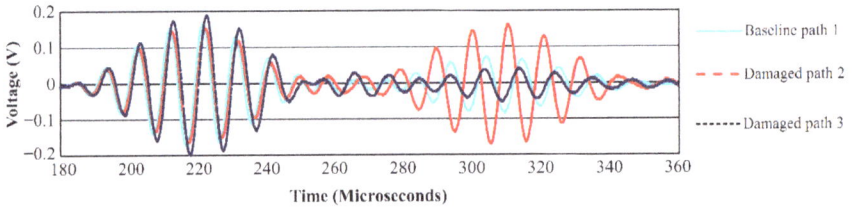

(b) lamp wave signals on fabric Plate (Baseline) path 1 and changing the damage Path 2& Path 3.

Figure 6.15: Lamb wave analysis for damage identification.

A detailed description of the time reversal method for damage detection processes is illustrated in Figure 6.16. As shown in the figure, a network of transducers is installed on the structure, and a signal is activated from one transducer and recorded at another. Each transducer has two functions as a sensor and as an actuator at the same time; therefore, piezoelectric transducers are generally used. According to Figure 6.16, the time reversal process begins with the actuated signal sent by piezoelectric transducer (A) and recorded at piezoelectric transducer (B). The signal received at piezoelectric transducer (B) is inverted over time (i.e., $V_B(t) \rightarrow V_B(-t)$) and then returned from piezoelectric transducer (B) to piezoelectric transducer (A), where it is recorded. Finally, the signal received at piezoelectric transducer (A) is time reversed and compared to the original signal.

Due to the importance of the time reversal method, the researchers focused on improving its performance. Wu et al. [313] detected bolt loosening in thermal protection system panels and modified the time reversal method to enhance the analysis performance of piezoelectric transducer responses by introducing a standard reemitted signal (SRS) (signal {4} in Figure 6.16), which passes through the damage (bolt loosening) again, as shown in Figure 6.17. When the SRS is resubmitted through the damage, it exhibits extensive patterns compared to the traditional time reversal method. Therefore, the SRS contains additional information about the damage (such as wave velocity), which can be extracted for damage range detection. Further details on recent studies aimed at improving the performance of the time reversal method are provided in Table 6.6.

To determine the location of the damages, the triangulation method [323], tomography techniques [324], the time inversion method [325], and AI methods [326] can be used. These methods require extensive calculations to train the network algorithm on how to locate damage based on damage data simulations. However, if the training conditions are altered, the algorithm may become ineffective. Damage locations can also depend on the propagation of energy from Lamb waves [327].

Table 6.6: Some recent studies on improving the performance of the time reversal method.

Method	Highlight	References
Wavelet	A wavelet-based signal processing technique was combined with the time reversal method to improve the time reversibility of Lamb waves in composite laminates.	[314]
SAMS	A modified time reversal method is proposed by using a single actuator and multiple sensors (SAMS). Therefore, the initial input signal and the secondary excitation signal are actuated by the same actuator.	[315]
VTRA	A modified time reversal method is proposed by using only one actuating–receiving step to obtain the reconstructed signal.	[316]
Optimization	The effects of the transducer-plate adhesive layer, parameters of excitation, size of transducer, and thickness of plate are investigated on time reversal method's performance to improve an effective strategy of damage detection.	[317]
Single-mode and two-mode lamp waves	The effects of single-mode and two-mode lamp waves on the activity of the time reversal method for the damage detection process are investigated both theoretically and experimentally.	[318]
ATROE	Improved performance of the time reversal method in damage detection by mitigating the effects of the time reversal operator.	[319, 320]
OTRP	Some different damage indices are proposed to improve the performance of the time reversal method for detecting damage.	[321, 322]

6.3 Electrical capacitance sensor technique-based structural health monitoring of composite structures

Most traditional nondestructive testing techniques, such as magnetic particle and ultrasonic testing, cannot monitor damage that begins during creep in composite structures,

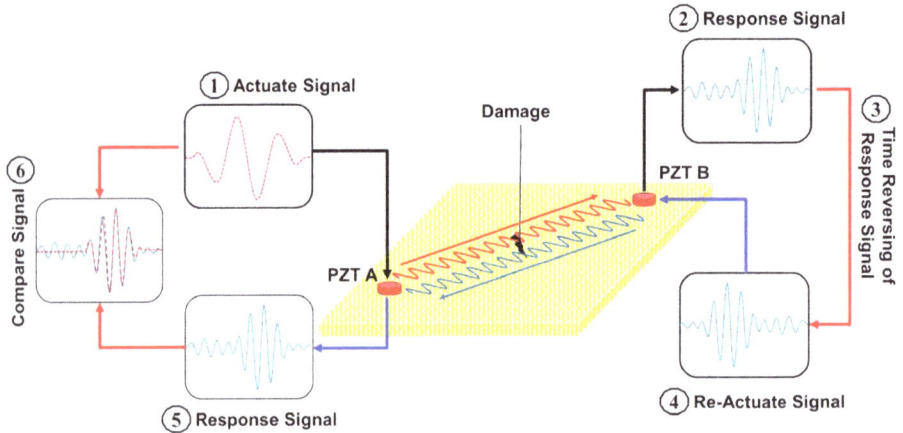

Figure 6.16: The time reversal method for damage detection processes.

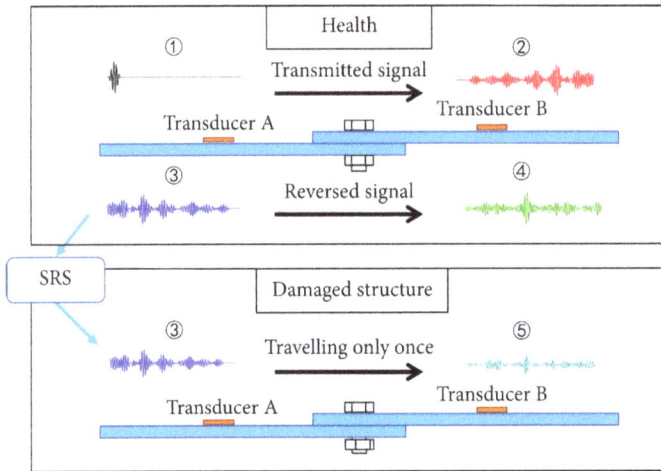

Figure 6.17: A schematic of the modified time reversal method [313].

because detecting damage before cracks form due to creep, for example, is a difficult task for these techniques to achieve. Furthermore, conventional sensing methods, which utilize strain gauges, vibration sensors, and piezoelectric-type sensors, as well as advanced sensors such as optical fibers, often require the installation of sensors either inside or outside the structure and are often expensive.

To address these drawbacks, alternative monitoring systems based on the intrinsic permittivity properties of materials have recently been investigated. The primary working principle of this type of sensor is to measure the electrical potential changes corresponding to material permittivity variations between arrays of electrodes mounted on the outer surface of the structure. Sensors that operate based on the dielectric properties of materials for damage detection offer several advantages, such as higher reliability and lower cost compared to other technologies, including conventional fiber-optic sensors. Furthermore, they can be used in large composite structures, unlike fiber-optic sensors, which cannot be applied in dense networks. Fiber-optic sensors also fail to detect cracks if the crack propagation does not intersect with the sensor. Additionally, the installation of fiber-optic sensors may lead to damage.

On the contrary, the electrical capacitance sensor offers several advantages, such as low cost, low sensitivity to noise, fast response, higher safety, and continuous operability under harsh environmental conditions.

The advantages of monitoring systems based on dielectric properties for damage detection in composite structures, compared to other techniques such as traditional fiber-optic sensors, are significant. Fiber-optic sensors often suffer from unsatisfactory reliability and high costs, making it impractical to deploy an extensive network of these sensors for large composite structures. Additionally, crack detection using fiber-optic sensors can fail if the crack propagation does not intersect with the sensors. Furthermore, the installation of thicker types of fiber-optic sensors may itself become a source of damage. Conversely, electrical capacitance sensors offer several advantages, including low cost, fast response, a non-noisy method, high safety, and continued operability under harsh environmental conditions.

6.3.1 History and literature behind the electrical capacitance sensor technique

In the 1980s, researchers at the US Department of Energy's Morgantown Energy Technology Center invented the electrical capacitance sensor for the first time and used it to measure fluidized bed systems. The technology has grown rapidly over the past 10 years and has become both popular and important for monitoring industrial processes due to its numerous advantages.

The electrical capacitance sensor is one of the most mature and promising new methods. It converts the permittivity of the piping system into interelectrode capacitance, which constitutes the electrical capacitance sensor's forward problem. The capacitance measuring circuit collects the capacitance data and transfers it to the imaging computer. The imaging computer reconstructs the distribution image using a suitable algorithm, referred to as the electrical capacitance sensor's inverse problem. This method is characterized by its low cost, fast response, and nonintrusive nature, with a broad range of applications and high level of safety. The need for more accurate measurements using electrical capacitance sensors has led to the study of factors that influence and affect the sensitivity and sensitive domain of electrical capacitance sensor electrodes. However, the main drawback of the electrical capacitance sensor method is that extensive FE calculations must be performed to obtain a sufficient number of sets of electric potential differences. This is the main drawback of the method identified so far.

Many fundamental studies have been reported using electrical capacitance sensors to detect various types of damage in composite and steel pipes. A 3D electrical capacitance sensor model for damage detection in carbon fiber-reinforced polymer composite pipes is studied by Zhao et al. [114]. They used the corresponding capacitance of permittivity changes in pipe material due to damage as an input of an open-loop system to identify the damage by checking the system transfer function stability based on the capacitance changes. Use electrical capacitance sensor for monitoring mechanical property degradation due to material water absorption over time in the Glass fiber reinforced epoxy composite pipes under internal hydrostatic pressure and thermal effect by Altabey et al. [120]. There is another related study by these authors, on monitoring a long-term creep thermomechanical fatigue behavior monitoring of BFRP composite pipeline using electrical capacitance sensors and deep learning algorithm [121]. They also used the electrical capacitance sensor in other related works and demonstrated the efficiency and accuracy of the sensor used in high high-accuracy detection of various damages such as fatigue cracks in BFRP pipes [119], delamination identification in BFRP pipes [117, 118], internal corrosion in steel pipes [35, 122], and monitored the nano-delamination embedded in composite nanopipes [113].

6.3.2 Electrical capacitance sensor modeling and simulation

The electrical capacitance system, which includes a sensor, a capacitance measuring circuit, and an imaging computer, is shown in Figure 6.18.

An electrical capacitance sensor consists of an insulating pipe, measurement electrodes, a radial screen, and an earthed screen. The measurement electrodes are mounted symmetrically around the circumference of the pipeline. The radial screen is fitted between the electrodes to block the electric field external to the sensor pipeline and reduce interelectrode capacitance. The earthed screen surrounds the measurement electrodes to shield them from external electromagnetic noise. The electrical capacitance sensor converts the permittivity of the inner media flow into interelectrode capacitance, which is referred to as the electrical capacitance sensor's forward problem. A capacitance measuring circuit collects the capacitance data and transfers it to an imaging computer. The imaging computer reconstructs the distribution image using a suitable algorithm, a process known as the electrical capacitance sensor's inverse problem. In most applications, the electrical capacitance sensor electrodes are mounted outside the pipeline, a configuration known as an external electrode electrical capacitance sensor.

The electrical capacitance sensor is composed of a composite pipe and features a ring of electrodes, which are separated from each other by small gaps. Each electrode covers a 30° angle around the pipe's outer surface. Figure 6.19 shows the cross-section of a 12-electrode electrical capacitance sensor system, where R_1 is the inner composite

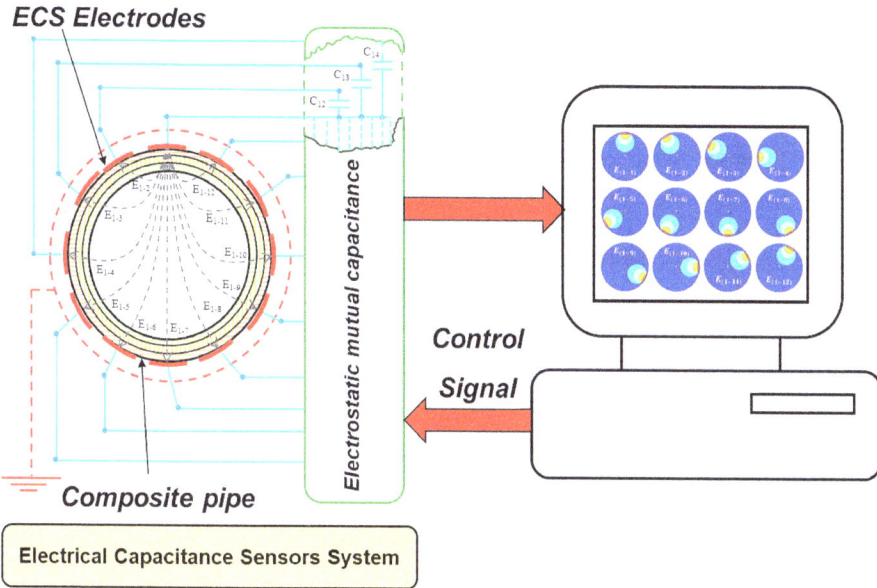

Figure 6.18: Sketch of an electrical capacitance sensor system.

pipe radius, R_2 is the outer composite pipe radius, R_3 is the earthed screen radius, and the radial screen is connected to the outer composite pipe. The dielectric permittivity of the composite material and the internal medium are denoted as ε_g and ε_w, respectively.

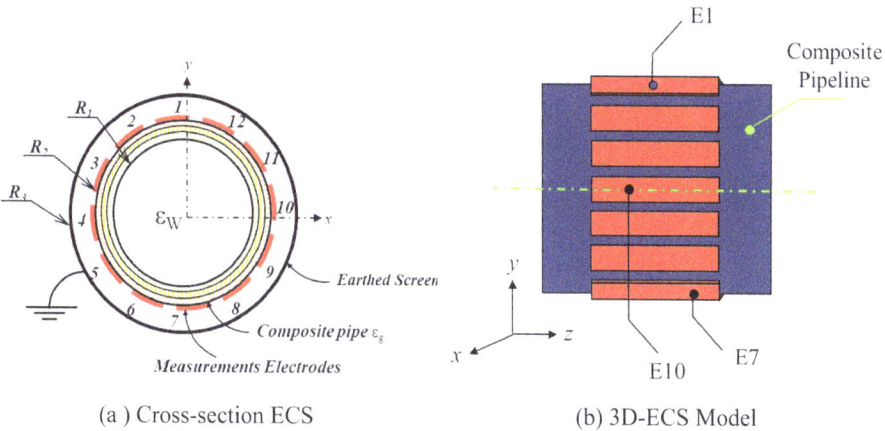

(a) Cross-section ECS   (b) 3D-ECS Model

Figure 6.19: Twelve-electrode electrical capacitance sensor system.

6.3.3 Electrical capacitance sensor electrode excitation strategy

Figure 6.20 introduces the electrode excitation strategy for the ECT working principle, referred to as the traditional excitation strategy of single-electrode excitation. In this strategy, all sensor electrodes function as detection electrodes and are grounded ($\varphi = 0$) except for one electrode that is excited ($\varphi = V_0$), for example, when $E1$ is excited, all other electrodes ($E2$–$E12$) in the network are grounded. Subsequently, the electrical capacitance sensor measurements of capacitances between the excited electrode and the other electrodes can be extracted one after the other, such as $E1$–$E2$, $E1$–$E3$, ..., $E1$–$E12$. Then, $E2$ is excited, and measurements are made between $E2$–$E3$, $E2$–$E4$, ..., $E2$–$E12$. In this strategy, the total number of independent capacitance measurements for a 12-electrode ECT sensor can be calculated using the formula $M = \frac{N(N-1)}{2} = 66$, where N is the number of sensor electrodes.

6.3.4 3D electrical capacitance sensor governing equation

The corresponding capacitance between electrode pairs C_{ij} can be computed between the ith electrode and the jth electrode as follows:

$$C_{ij} = \frac{Q_{ij}}{\Delta V_{ij}}, \quad (\Delta V_{ij} = V_i - V_j) \tag{6.4}$$

where Q_{ij} is the charge induced on the electrode j while the electrode i is excited with a voltage, V_{ij} is the EPD between electrode pairs i and j. And, C_{ij} in the linear model of an electrical capacitance sensor can be expressed as follows:

$$C_{ij} = S_{ij}(x, y, z)\varepsilon(x, y, z) \tag{6.5}$$

where $S_{ij}(x, y, z)$ is the sensitivity matrix, and $\varepsilon(x, y, z)$ is the permittivity distribution. The potential distribution $\varphi(x, y, z)$ inside the electrical capacitance sensor can be determined using Poisson's equation as follows:

$$\nabla.\varepsilon(x, y, z)\nabla\varphi(x, y, z) = 0 \tag{6.6}$$

where $\nabla.\varepsilon(x, y, z)$ is the permittivity distribution, and $\nabla\varphi(x, y, z)$ is the gradient of the potential distribution. Finally, the elements of the 3D electrical capacitance sensor sensitivity matrix over $p(x, y, z)$ volume can be calculated as follows:

$$S_{ij}(x, y, z) = -\iint_{p(x,y,z)} \frac{\nabla\varphi_i(x, y, z)}{V_i} \cdot \frac{\nabla\varphi_j(x, y, z)}{V_j} \, dxdydz \tag{6.7}$$

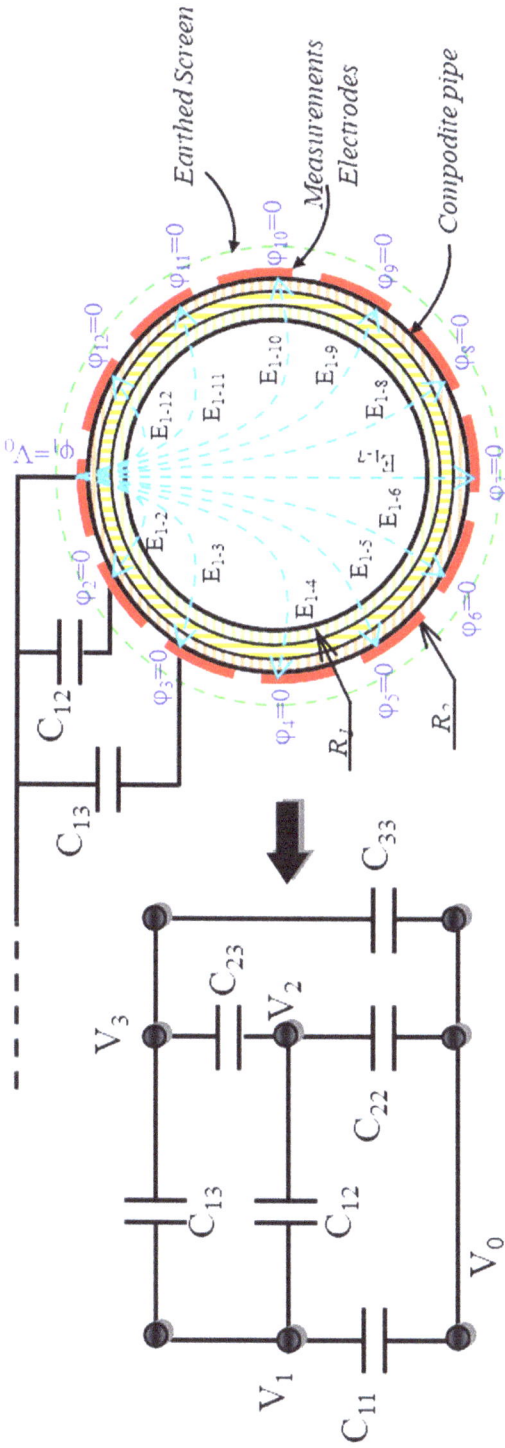

Figure 6.20: Structure of sensors with excitation strategy parameters.

6.3.5 Factors affecting the electrical capacitance sensor technique

To use the electrical capacitance sensor more accurately, several works have been conducted to examine the parameters that influence its sensitivity. Three basic parameters affecting sensitivity are: (1) the material of the pipe, (2) the permittivity of the internal medium within the pipe, and (3) the ratio of the pipe's thickness to its diameter. Recently, the author of this book demonstrated that the environmental temperature surrounding the electrical capacitance sensor is a fifth parameter that impacts its sensitivity.

6.3.6 Effect of the number of electrodes on the performance of electrical capacitance sensors

The effect of the number of electrodes in the electrical capacitance sensor for damage detection of composite structures has been reported by Altabey [118]. He used 8-electrode, 12-electrode, and 16-electrode configurations for damage detection in composite structures. The results presented are in excellent agreement with solutions available in the literature, thereby validating the accuracy and reliability of the proposed technique. Response surfaces were utilized to solve the inverse problems. He found that the sensitivity of the electrical capacitance sensor for damage monitoring depends on the number of electrodes mounted on the outer surface of the composite structures. As the number of electrodes increases, the sensitivity of the electrical capacitance sensor also increases, as shown in Figure 6.21. The performance of damage detection improves with an increasing number of electrodes mounted on the outer surface of the composite structures, with a high correlation factor of detection results near 1.0 for the 8-electrode, 12-electrode, and 16-electrode configurations. The error margin for damage detection depends on the number of electrodes mounted on the outer surface of the composite structures. A larger number of electrodes are required to maintain high estimation performance for identification, as shown in Figure 6.22.

6.3.7 The life of electrical capacitance sensor

The electrical capacitance sensor functions like any capacitor. It has two aspects of life: the first one is shelf life, when capacitor is stored for a long time without electric charge and leakage current, and resistance may increase and capacitance may decrease. However, these changes are typically small after the storage around 2 years at room temperature for general capacitors, or around 6 months for low-leakage products, and are not considered a practical issue. These changes are attributed to chemical reactions between the electrolyte and the dielectric oxide film. So one of the reasons why leakage current increases is if capacitor is exposed to high-temperature

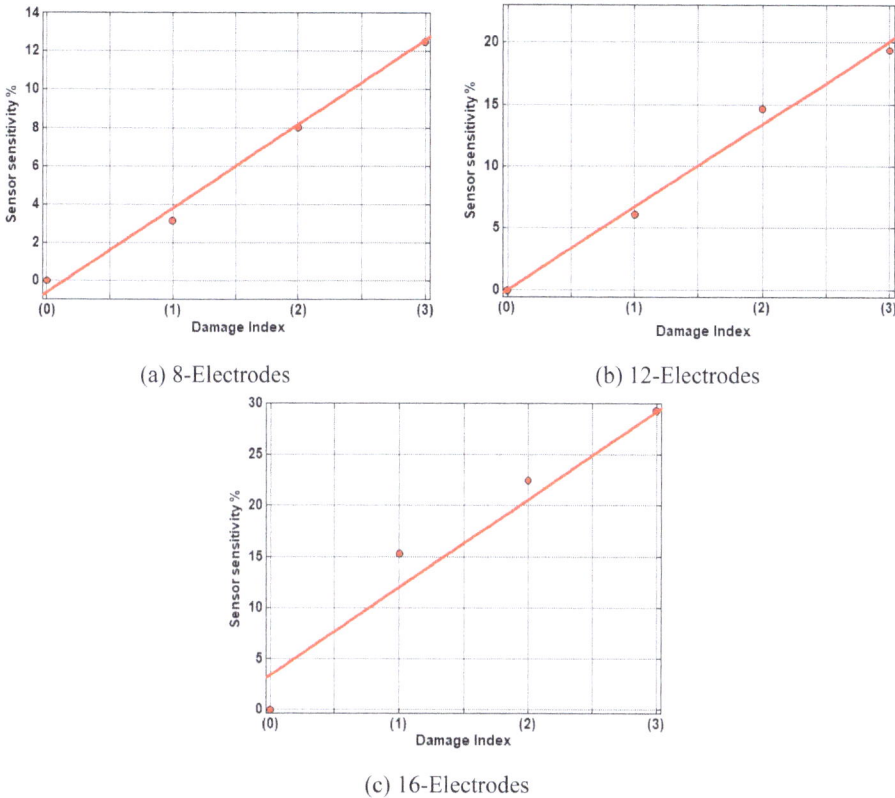

(a) 8-Electrodes

(b) 12-Electrodes

(c) 16-Electrodes

Figure 6.21: Electrical capacitance sensor sensitivity versus damage index.

atmosphere, which may bring changes in characteristics. The second one is load life, when capacitor is applied with DC voltage for a long time, and capacitance will reduce. Leakage current generally remains low because the dielectric oxide film is always repaired by the applied DC voltage, consuming electrolyte. Changes in capacitance are primarily caused by the loss of electrolyte through dissipation and decomposition, processes that are accelerated in high-temperature environments. General changes of each characteristic under shelf life and load life testing at 85 °C of electrical capacitance sensor are shown in Figure 6.23.

6.4 Fiber-optic sensor technique-based structural health monitoring of composite structures

Sensing technology is one of the fastest-growing high-tech industries in the world today. The new sensor not only aims for high precision, wide range, high reliability, low power consumption, and miniaturization but also evolves toward integration,

(a) 8-Electrodes

(b) 12-Electrodes

(c) 16-Electrodes

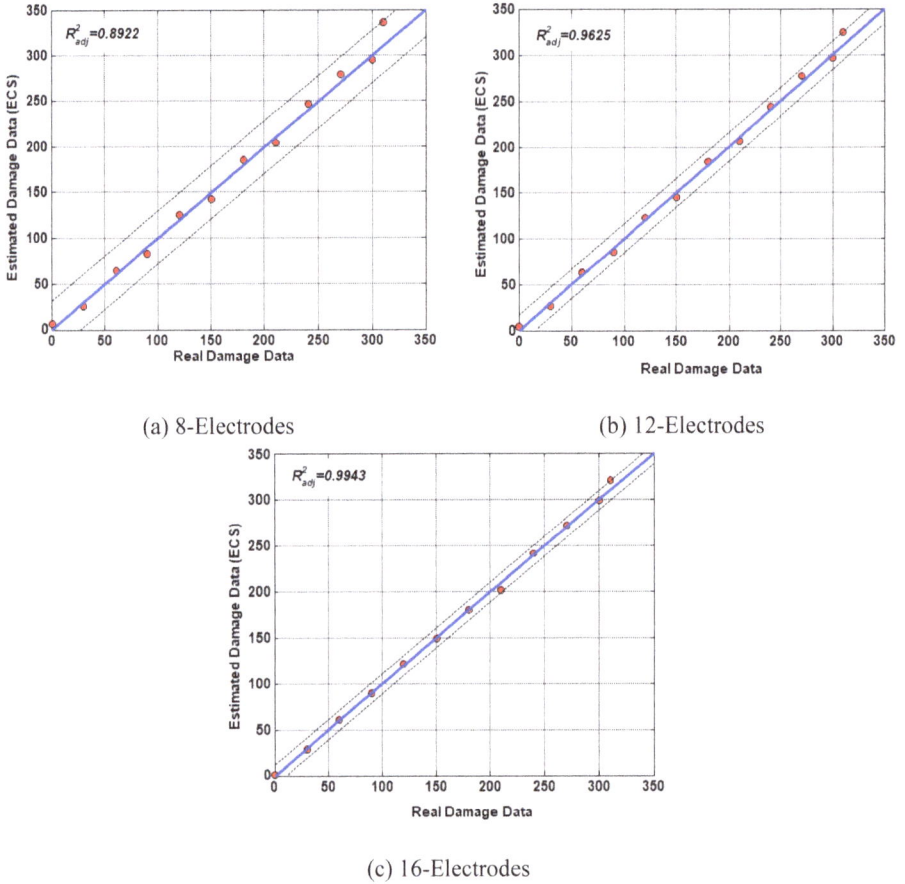

Figure 6.22: A comparison between estimated damage data (electrical capacitance sensor data) and real data in composite structures.

multifunctionality, intelligence, and networking to meet the needs of various fields such as industry, agriculture, national defense, and scientific research.

Numerous studies have been conducted on pipelines using fiber-optic sensors to estimate the effects of pipeline damage through both experimental and numerical prediction methods. Nikles and Briffod [328] introduced a technique to address the impact of blockages in hydrocarbon pipelines using a fiber Bragg grating sensing system, which provides distributed sensing capabilities. Their proposed approach yielded results that simulated the effects of pipeline blockages, thereby validating the effectiveness of their introduced technique. Inaudi and Glisic [329] reported several important field application examples of fiber-optic sensing with the ability to measure temperature and strain at thousands of points using a single fiber optic. Their approach demonstrated important applications for monitoring slender pipelines installed in oil

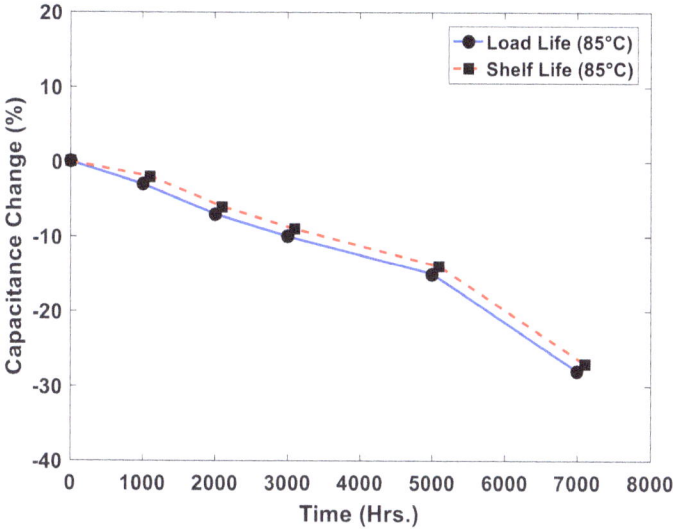

Figure 6.23: General changes of electrical capacitance sensor characteristics under shelf life and load life.

wells and coiled tubing. Their approach could detect pipeline leaks and prevent the failure of pipelines installed in refineries, and for detecting the hot spots in high-power cables. Meinert et al. [330] proposed a method for detecting and preventing serious damage to pipelines, mainly caused by interference of several noise sources. They showed that their permanent monitoring semi-intelligent system could reduce the need for online inspections. Yan and Chyan [331] presented the results of simulation studies and theoretical discussions on suppressing unfavorable fiber-optic non-linearity and stimulated Brillouin scattering by using a statistical approach. They also demonstrated that because there are several stimulated Brillouin scattering threshold factors, there is a high possibility that multiple strong interactions exist among these factors. They further showed that conventional methods of plugging each factor one at a time are not efficient as interactions between factors are ignored. Their results illustrated stimulated Brillouin scattering suppression by 11 dB to achieve the maximum sensing distance of up to 110 km. Besides, they employed a 3-bit simplex coding when launching the laser pulses to enhance the SNR at the receiver. Simulation results demonstrated a 1.38 gain in SNR and a higher stimulated Brillouin scattering threshold.

For the development of non-slippage fiber optics, the bonding and point fixation methods were investigated experimentally, and the critical effective sensing length for long-gauge fibers was studied [329]. A combination of an artificial neural network and a distributed fiber-optic vibration sensing system based on long-distance fiber has been used to collect the vibration signal responses of soil around the pipeline [332–335]. Additionally, the technology of the distributed fiber-optic vibration sensing system, based on a phase-sensitive optical time-domain reflectometer, was utilized to study the hydrostatic leak test of water pipelines [114] and to identify fatigue damage in composite pipeline systems using electrical capacitance sensors [336].

6.4.1 History and literature behind the fiber-optic technique

Once fiber-optic sensors emerged in the mid-1970s, they were highly valued by research departments in various countries. The United States was the first country to conduct research on fiber-optic sensors and achieved the highest level of advancement in the field, and its progress in both military and civilian applications is very rapid. In military applications, their research and development mainly include fiber-optic sensors for underwater detection, aviation monitoring, gyroscopes, and nuclear radiation detection. In the mechanical and civil engineering fields, fiber-optic sensors are mainly used to monitor important parameters such as current, voltage, and temperature in power systems, as well as to track stress changes in bridges, pipelines, and other important buildings.

6.4.2 Common types of fiber-optic sensors

Selecting the correct fiber-optic sensor type to monitor the excitation forces applied to a structure is important for detection. Before delving into the details of the working principles of fiber optics, a concise review of common types of fiber-optic sensors is provided. Among the common types of fiber-optic sensors, point sensors and distributed sensors are of interest in the pipeline monitoring field. Figure 6.24 presents the classification of fiber-optic sensor categories based on guiding optical principles, including distributed sensors for distributed sensing in large structures, short-gauge sensors for point-sensing in homogeneous materials such as steel, and long-gauge sensors for point-sensing in heterogeneous materials such as concrete.

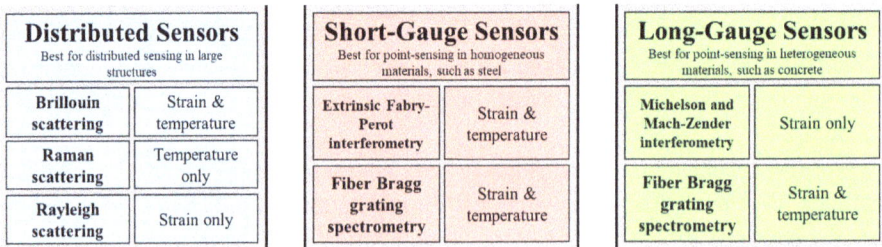

Distributed Sensors Best for distributed sensing in large structures		Short-Gauge Sensors Best for point-sensing in homogeneous materials, such as steel		Long-Gauge Sensors Best for point-sensing in heterogeneous materials, such as concrete	
Brillouin scattering	Strain & temperature	Extrinsic Fabry-Perot interferometry	Strain & temperature	Michelson and Mach-Zender interferometry	Strain only
Raman scattering	Temperature only	Fiber Bragg grating spectrometry	Strain & temperature	Fiber Bragg grating spectrometry	Strain & temperature
Rayleigh scattering	Strain only				

Figure 6.24: Schematic diagram of the classification of fiber-optic sensors based on application type.

6.4.3 Description of fiber Bragg grating sensors

The fiber Bragg grating system consists of an article interrogator projecting infrared light into the core of an optical fiber. As a white color, broadband light travels along the fiber; it passes through grating segments, also known as fiber Bragg grating, which acts as a series of article filters. These grating segments can filter out certain wavelengths or colors while allowing others to pass.

This occurs by periodically altering the refractive index of the fiber, which determines which wavelengths can pass through and which will be reflected. External factors, such as heat and vibration, cause the reflected light to shift in wavelength. These variations can then be converted into physical engineering units, such as temperature and strain. An example of fiber Bragg grating detection technology is shown in Figure 6.25.

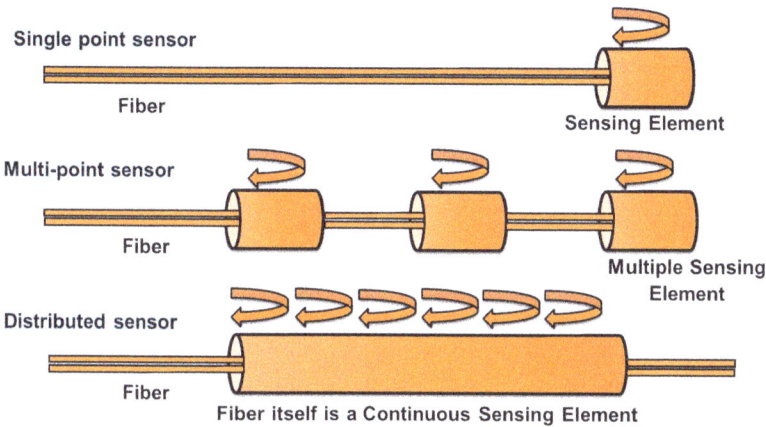

Figure 6.25: Schematic diagram of three fiber Bragg grating sensors.

6.4.4 Working principle of fiber Bragg grating sensors

The distributed fiber Bragg grating detection method uses fiber Bragg grating sensors to detect vibration signals along a pipeline. When events such as human activities or mechanical operations occur near the sensing optical cable, the vibration signals generated by these events cause the optical cable to strain, resulting in the phase of the light in the optical cable. Consequently, the polarization state of the light changes, and the system recognizes and alarms the detected changes (see Figure 6.26). As illustrated in Figure 6.26, the fiber Bragg grating sensor system integrates a broadband light source, fiber Bragg gratings, a wavelength interrogator, and system software.

When broadband light is projected onto a fiber Bragg grating, reflection occurs at the grating. Some light having wavelengths that satisfy the Bragg condition of Equation (6.8) is reflected, and the others pass through the grating:

$$\lambda_B = 2n_e A \tag{6.8}$$

where λ_B is the Bragg wavelength, n_e is the effective refractive index, and A is the grating period. When strain is applied to a fiber Bragg grating, a proportional shift in the Bragg wavelength is expected to occur. The strain can be easily determined by analyzing the change in the wavelength. According to this principle, fiber Bragg grating sensors can detect the grating period change due to strain variation, and they can measure strain without the influence of noise and light intensity perturbations. The wavelength shift is proportional to the strain, and absolute strain can be measured as follows:

$$\frac{\Delta \lambda_B}{\lambda_B} = \left\{ 1 - \frac{n_e^2}{2} \left[P_{12} - \upsilon (P_{11} + P_{12}) \right] \right\} \varepsilon_B \tag{6.9}$$

where P_{ij} are the silica photoelastic tensor components, ε_B is the strain of the fiber grating, and υ is the Poisson's ratio.

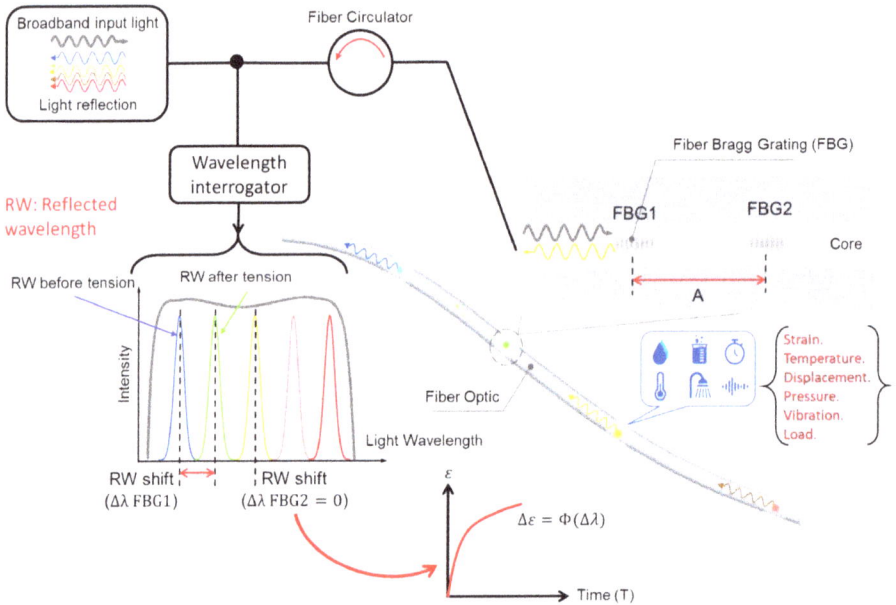

Figure 6.26: Schematic of the working principle of fiber Bragg grating sensors.

6.4.5 Improve the design of the fiber Bragg grating for large-strain sensor

A novel type of large-strain sensor has been developed based on pre-relaxation and continuous sensing technology. This design aims to enable comprehensive strain monitoring throughout the entire process of prestressed structures. This new type of large-strain sensor was tested on nine large-strain sensor specimens in research laboratories at the National and Local Joint Engineering Research Center for Basalt Fiber Production and Application Technology, Southeast University, Nanjing, Jiangsu, China. It has proven its efficacy in capturing the strain relevant to the post-tensioned operational state of the structures. Additionally, a sensing performance of up to 18,923 $\mu\varepsilon$ can be facilitated and further enhanced through the pre-relaxation and continuous sensing technique [337].

A large-strain sensor, integrating fiber Bragg grating's pre-relaxation and continuous monitoring technologies, was introduced, as illustrated in Figure 6.27. This sensor combines a straight fiber Bragg grating (S-FBG) and a pre-relaxed fiber Bragg grating (R-FBG) within a single measurement section. As loading commences and the FRP plate begins to stretch, the S-FBG starts its operation. Concurrently, the R-FBG initiates its straightening process. Once the S-FBG approaches failure, the R-FBG, having fully straightened, assumes the operational role.

The design schematic of the large-strain sensor is illustrated in Figure 6.27. Assuming that the length of the sensor's measurement section (or the gauge length) is C_o, the pre-relaxation degree of the sensor is β, then the pre-relaxation length of the sensor can be determined.

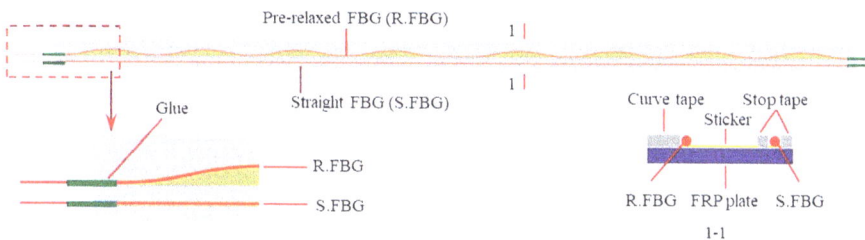

Figure 6.27: Schematic of the large-strain sensor utilizing pre-relaxation and continuous monitoring technology.

Represented as βC_o. Thus, the actual length of the sensor L_o will be:

$$L_o = C_o + \beta C_o \tag{6.10}$$

Let L_c be the chord length of a single wave, L_a be the length of a single arc, n be the section number, L_m be the horizontal length in the middle, and L_e be the horizontal length at the end.

Based on the geometric relationship, the expressions for C_o and L_o are as follows:

$$C_o = 2L_c.n + L_m + 2L_e \tag{6.11}$$

$$L_o = 2L_a.n + L_m + 2L_e \tag{6.12}$$

From formulas (6.10), (6.11), and (6.12), L_c and L_a can be obtained as follows:

$$L_c = \frac{C_o - L_m - 2L_e}{2n} \tag{6.13}$$

$$L_a = \frac{L_o - L_m - 2L_e}{2n} \tag{6.14}$$

As depicted in Figure 6.28, given the known values of L_c and L_a, we can calculate the central angle $2a$, radius r, and chord height h as follows:

From the equation $2a = L_a/r$, we can deduce:

$$L_a = 2ra \tag{6.15}$$

From the equation $= L_c/2$, we can deduce:

$$L_c = 2r \sin a \tag{6.16}$$

By dividing both sides of formulas (6.15) and (6.16), we obtain the following:

$$\frac{\sin a}{a} = \frac{L_c}{L_a} \tag{6.17}$$

Using Matlab and employing the dichotomy method, we can calculate a.

Based on formula (6.15) and the Pythagorean theorem, we can determine the values for r and h as follows:

$$r = \frac{L_a}{2a} \tag{6.18}$$

$$h = r - \frac{\sqrt{4r^2 - c^2}}{2} \tag{6.19}$$

6.5 Artificial intelligence

In this section, we complete the review of the third subsystem of structural health monitoring system components: the health assessment system. An artificial intelligence system that can serve the role of diagnostic algorithms and information management to assess details about their origins would prove to be immensely beneficial.

Artificial intelligence aims to mimic human intelligence by developing computer programs to solve complex problems. In early applications, the framework of artificial intelligence-based schemes for structural health monitoring and damage detection of composite structures is presented in Figure 6.29.

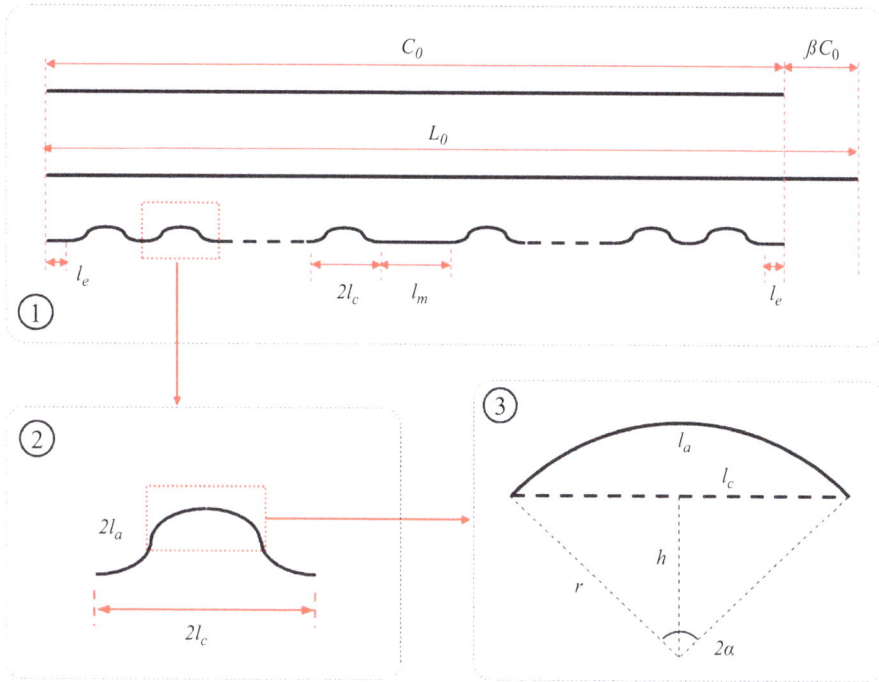

Figure 6.28: Design diagram of the R-FBG sensor.

6.5.1 Machine learning

Although the programs developed by artificial intelligence are based on human knowledge, they have surpassed human capabilities in many cases, such as playing chess [338]. This is considered the most confrontational challenge experienced by most artificial intelligence systems so far [339], for which the machine learning concept was developed to address this challenge. The way the machine learning algorithm is designed allows the program to extract the required information from data to systematically learn and accomplish a given task [340]. To achieve this goal, preprocessing of data is required for feature extraction and characterization in terms of quality. This step is called "feature extraction" [341]. Feature extraction is used to train the machine learning system to recognize different patterns in the data. Figure 6.30 illustrates the procedure for training a machine learning algorithm. Many researchers have studied machine learning technologies in nondestructive testing applications. Table 6.7 provides some examples of these studies.

Artificial Intelligence (General Definition): The theory and development of computer systems able to perform tasks that normally require human intelligence, such as visual perception, speech recognition, decision-making, and translation between languages.

Simulation of Human Intelligence by Machines.

SHM / Big Data

Data / Information

Learn and Solve Problem !

Algorithm

- Sensing and context of data collection and generation
- Data storage and representation
- Data analysis, science and engineering
- Data interference and visualization
- Computing support for data analysis
- Data intensive applications

- Clustering algorithms vs. linear regression
- Optimization algorithm (Genetic algorithm)
- Bayesian updating
- Fuzzy logic
- Decision trees
- Artificial neural network
- Support vector machine
- Machine learning
- Deep learning

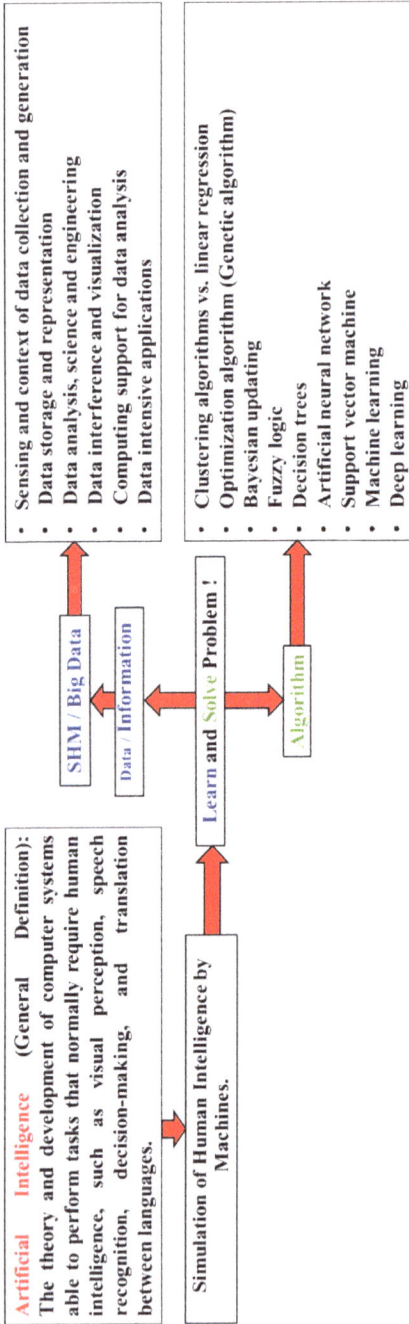

Figure 6.29: The framework of artificial intelligence-based schemes for structural health monitoring.

- **Designing Input-output Scheme**

- **Vibration signal Aquistion**

- **Pre-procissing of raw signal**

- **Extructtion machine learning Input from processed signal**

- **Selection of machine learning model**

- **machine learning implementation**

- **Post-procissing of machine learning output**

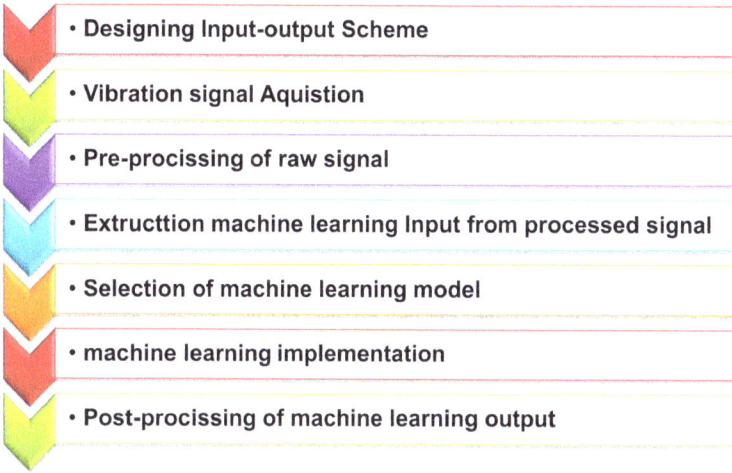

Figure 6.30: Machine learning algorithm training procedures.

Table 6.7: Some recent studies on the application of machine learning algorithms in NDE.

Algorithm	Nondestructive testing methodology	References
K-means algorithm	Acoustic emission signal	[342, 343]
	Thermal imaging	[344]
	Pulse eddy currents	[345]
	Thermography	[346, 347]
DBSCAN	Ultrasonic testing	[348]
	Ultrasonic Lamb wave	[349, 347]
	Impact echo, ultrasonic pulse echo	[350]
	Laser ultrasound	[351]
Spectral clustering	Terahertz spectroscopy	[352]
	Vibration signals	[353]
	Spectral kurtosis	[354]
Hierarchical clustering	Ultrasonic echo testing	[355]
	Electromechanical impedance method	[354, 356]
Association analysis	Fiber-optic sensors	[343]
	Multipoint laser vibrometers	
	Acoustic emission sensors	
Support vector machine	X-ray casting	[357]
	Long-range ultrasonic testing	[358]
	Raman spectroscopy	[359]
K-nearest neighbor	Microwave testing	[360]

Table 6.7 (continued)

Algorithm	Nondestructive testing methodology	References
Neural networks	Ultrasonic pulse velocity test	[361]
	Thermograms	[362]

6.5.2 Deep learning

As shown in Figure 6.30, the machine learning algorithm's data processing depends on the strength of the extracted features in representing the data. Therefore, it is essential to extract optimal features that can effectively describe the characteristics of the input data, and must always identify the optimal features manually for training; however, this approach is not practical. To reduce the complex dependence of machine learning applications on handcrafted features, deep learning algorithms, such as deep neural networks, were developed. Deep learning algorithms are considered a special type of machine learning algorithm because they have the ability to extract optimal features directly from raw datasets without user intervention. Deep learning systems are programmed to establish a direct mapping from raw datasets to targets without the need to extract features a priori [363]. By learning to hierarchically extract high-level and abstract features from learned low-level and simple features [354], deep learning can effectively handle complex problems [365–367]. Figure 6.31 illustrates the procedure for training a deep learning algorithm.

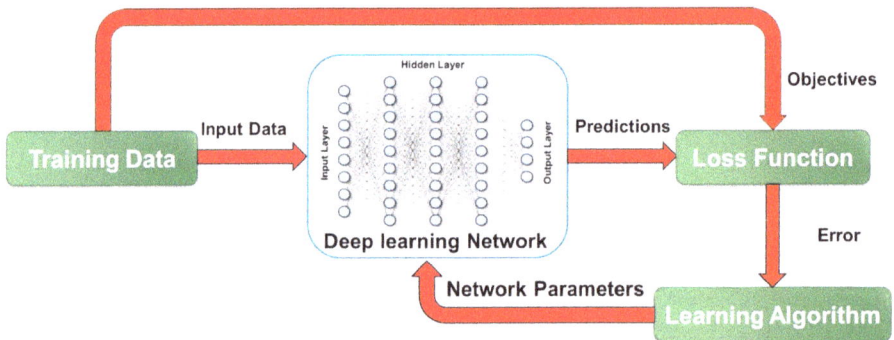

Figure 6.31: Training procedures of deep learning algorithm.

6.5.3 Artificial neural networks

The artificial neural network is one of the techniques of artificial intelligence, alongside expert systems, genetic algorithms, fuzzy logic, machine learning, and Bayesian networks. Briefly, artificial neural networks simulate the functions of a biological brain and have been widely used in various research fields, such as computer science [368, 369], medicine [370, 371], finance [372, 373], social sciences [374, 375], and engineering [376, 377]. They have been applied in several areas, including speech recognition [378–385], the diagnosis of hepatitis [386], the recovery of telecommunications from faulty software [387], image recognition [388–390], and failure detection in laminated composite materials [391–400]. Their ability to model nonlinear problems and their robustness in noisy environments make them the preferred choice for these types of applications [327, 401, 402].

Artificial neural networks are able to extract patterns and detect trends from complex or imprecise data that are too complicated to be analyzed by other diagnostic techniques.

It is clear that artificial neural network models possess several inherent properties that distinguish them from traditional computational models. They are composed of massively parallel networks of simpler computational elements, which can enhance inherent parallel computation rates and provide a higher degree of robustness and fault tolerance. Given a system and a set of "training" patterns, the neural network can adaptively "learn" about the system by adjusting its internal parameters to align with the training examples.

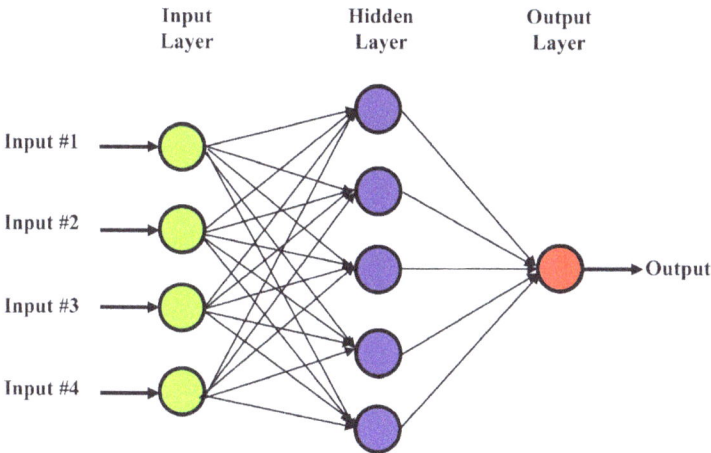

Figure 6.32: A simple model of a three-layer artificial neural network (input, hidden, and output).

There are various architectures, training procedures, and testing procedures for artificial neural networks. Generally, an artificial neural network consists of a potentially large number of nodes or neurons, which serve as processing elements. A neuron influences the behavior of other neurons through a weight. Each neuron transfers the result of its computations to its outgoing connections, which calculate a nonlinear weighted sum of its inputs [403]. The behavior of the network is largely based on the interactions between these neurons. As shown in Figure 6.32, the network is composed of several layers of neurons: input layers, hidden layers, and output layers. The input layer functions to transfer the input data and distribute it to the hidden layers, which are not directly accessible to the user. The hidden layers perform all the necessary computations and pass the results to the output layer, which provides the final output to the user.

6.5.4 Damage identification with artificial neural networks in composite structures

The real reason for using artificial neural networks in structural health monitoring is their ability to identify various types of damage in different locations within the same structure, making damage detection a complex process.

6.5.4.1 Gray-box model

White-box models are fully derived from first principles, that is, physical, chemical, or biological laws and have no dependence on measured data. Black-box models, on the other hand, are solely based on measured data, with little prior knowledge of the system's dynamics. Gray-box models (Figure 6.33) are a combination of white-box and black-box models, utilizing both first principles and the information contained in measured data. In gray-box models, the structure is determined by first principles, while the parameters are estimated from the measured data.

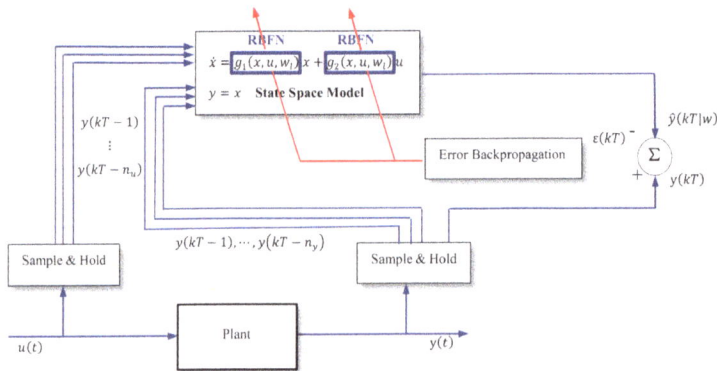

Figure 6.33: System identification using gray-box model.

Parametric system identification approaches are commonly used but require a priori knowledge of restoring force characteristics. Nonparametric approaches do not require a priori information; however, they typically lack direct associations between the model and the system dynamics.

When artificial neural networks are implemented using the "black-box" approach, limited system information can be obtained through traditional techniques. The intelligent parameter varying (IPV) method combines the advantages of both traditional parametric and nonparametric approaches. It utilizes an embedded radial basis function as the activation mechanism for neurons to estimate the constitutive characteristics of inelastic and hysteretic restoring forces in a multi-degree-of-freedom structural system.

The IPV technique, that is, a gray-box approach that combines the advantages of "white-box" and "black-box" methodologies, has been developed. In this approach, the model structure can be determined using first principles, while the nonlinear adaptation and learning capabilities of artificial neural networks are utilized to identify the nonlinear, time-varying system's dynamics that would be difficult to model and identify using traditional "white-box" and "black-box" methods.

A nonlinear system with full-state measurement is represented by the linear parameter varying model structure:

$$\dot{x} = f_1(x, u) \cdot x + f_2(x, u) \cdot u \tag{6.20}$$

$$y = x \tag{6.21}$$

The IPV approach preserves the structural model inherent in the system without requiring a priori representations of the nonlinearity $f_1(x, u)$ and $f_2(x, u)$, and these terms are represented by separate artificial neural networks $g_1(x, u, w_1)$ and $g_2(x, u, w_2)$ as follows:

$$\dot{x} = g_1(x, u, w_1) \cdot x + g_2(x, u, w_2) \cdot u \tag{6.22}$$

$$y = x \tag{6.23}$$

The structural model is preserved by incorporating an artificial neural network into it, thereby retaining a portion of the information from the original structural model. To maintain the association between system dynamics and the construction of the artificial neural network, the IPV approach is utilized for structural health monitoring and damage identification.

Zhao et al. [288–290] developed a system identification framework for structural hysteresis based on a Bayesian updating approach. This was achieved by collecting between IPV of a "gray-box" model with an artificial neural network approach, a genetic algorithm, and a transitional Markov chain Monte Carlo method. Recent studies on the application of artificial intelligence algorithms in the structural health monitoring of composite structures are summarized in Table 6.8.

Table 6.8: Some recent studies on the application of artificial intelligence algorithms in structural health monitoring of composite structures.

References	Method	Highlight	Model
[391, 392]	Feed-forward neural network	The two different types of damage (delamination and stiffness reduction) in glass fiber-reinforced polymer due to transverse cracks or impact damage was detected. Analytical models were used for training the network to predict the dynamic behavior of the structure taking into account various damage scenarios by main neural network and two subnet neural networks. The DSD method was used for increasing the speed of convergence neural network learning.	Glass fiber-reinforced polymer beam
[393, 394]	Artificial neural network and genetic algorithm	The delamination location in a multilayer glass fiber-reinforced polymer beam was detected. Two different procedures were followed (the FE model and locating errors, envisioned a set of possible damage scenarios). The finite element model was used for training the network with natural frequencies of the beam.	Multilayer glass fiber-reinforced polymer beam
[395]	Generalized regression neural network, linear neural networks, backpropagation neural network, radial basis function network	Four types of damage (circular holes, delaminations, linear surface cracks, and linear transverse cracks) in carbon fiber-reinforced polymer composite cantilever were identified.	Carbon fiber/Polestar composite cantilever
[396]	Backpropagation neural network and radial basis function network	The delamination of carbon fiber-reinforced polymer beams was predicted using natural frequencies as inputs to the network. A finite element model was used for training the network with different sizes and locations of delamination.	Carbon fiber/Polestar composite beam

[397]	Nonlinear artificial neural network	The crack between the fiber and the matrix in a unidirectional metal–matrix composite was detected. The 3D finite element model was used to generate the dataset with the plane stress and its damage variable for training the network.	Unidirectional metal–matrix composite
[398]	Artificial neural network	The size, shape, and location of delaminations in a fiber-reinforced plastic plate were detected using natural frequencies as inputs to the network. A 3D finite element model was used to generate the dataset for training the network.	Fiber-reinforced plastic laminated composite plates
[399]	Feed-forward neural network	The damage of a simply supported laminated composite plate using data from piezoelectric sensors as inputs of network was identified. The finite element model was used for training the network to obtain the necessary electrical potential on the sensors.	Laminated composite plates
[400]	Artificial neural network	The delamination behavior during drilling into carbon fiber-reinforced polymer plates due to high-speed drilling was studied. The EBPTA method was used for training the network.	Carbon fiber/Polestar laminate composite plate
[55]	Feed-forward neural network and radial basis function network	The fatigue life of a laminated composite plate under spectrum loading was predicted. The WPDD method was applied to evaluate various stress ratios and the residual strength effects caused by spectrum loading.	Carbon fiber/epoxy laminate composite plate
[404]	Artificial neural network and fault tree	The reliability evaluation of a laminated composite plate subjected to hydrostatic pressure was studied. An artificial neural network and a fault tree analysis were used to evaluate the reliability model.	Basalt fiber/Polestar laminate composite plate
[380]	Deep learning	The basalt fiber-reinforced polymer composite pipeline data extracted from long-gauge distributed fiber Bragg grating sensors were analyzed.	Basalt fiber-reinforced polymer composite pipeline

(continued)

Table 6.8 (continued)

References	Method	Highlight	Model
[405]	Deep learning	A damage assessment algorithm for composite sandwich structures was developed. The vibration mode shapes of the structures were used for learning the deep learning model.	Composite sandwich structures
[406]	Deep learning	Deep learning was exploited for the quantitative assessment of the visual detectability of various types of damage in laminated composite structures during service.	Laminated composite structures, such as those used in aircraft and wind turbine blades
[407]	Deep learning	Deep learning was used for damage labeling in a pin-joint composite truss structure. A finite element model was used for generating the data for training the deep learning model. The data acquisition problem was employed.	A pin-joint composite truss structure
[408]	Artificial neural network	The rapid convergence speed of artificial neural networks' gradient descent (GD) techniques and the global search ability of evolutionary algorithms (EAs) were exploited for network training.	Laminated composite structures
[409]	Artificial neural network	A newly modified damage indicator combined with an artificial neural network was proposed. LFCR was improved through a transmissibility technique.	Laminated composite structures
[410]	Machine learning	The possibility of detecting damage was demonstrated through the monitoring of acoustic emission signals generated in minicomposites with elastically similar components.	Unidirectional SiC/SiC composites

Chapter 7
Case studies on structural health monitoring of composite structures

7.1 Structural health monitoring of composite pipelines

A pipeline is an essential component of energy transportation systems. Pipelines can carry liquids, gases, or fine solids over long distances from their sources to the ultimate consumers via lifting stations that consist of variable speed pumps, valves, and control devices. These components cause changes in the operating conditions inside the pipelines, which leads to pipeline damage. It is vital to ensure the reliability and operational state of pipelines to maintain high performance of overall energy transportation systems. It is considered that numerical, theoretical, and/or experimental damage detection methods for pipelines require a large amount of labeled sample data.

7.1.1 Case study (1): predicting water absorption in composite pipes using electrical capacitance sensors integrated with deep learning approach

Monitoring the mass of liquid absorption in laminated composite structures, which have direct contact surfaces with working liquids, such as pipes, is crucial to prevent the sudden collapse of these structures due to degradation in strength and mechanical properties over time. An electrical capacitance sensor technique has been applied to monitor the mass of liquid absorption over time in laminated composite pipelines by measuring changes in the dielectric characteristics of the composite pipelines subjected to internal hydrostatic pressure from water and thermal effects. Results indicate that this technique is very effective. However, a significant challenge in utilizing this technique is its time-consuming nature, making it less practical for continuous monitoring. . In addition, the detection efforts required to calculate the mass of liquid absorption incur high costs and further loss of time. In this case study, a deep neural network model is employed to estimate the mass of liquid absorption in glass fiber-reinforced epoxy laminated composite pipelines by extracting features from datasets obtained through experimental and numerical measurements of the electrical capacitance sensor. The experimental and numerical data used to train and test the new deep neural network model were collected from the literature and the finite element model of the electrical capacitance sensor system, respectively. The results demonstrate excellent agreement between the finite element model data, available experimental data, and predictions made by the deep neural network, with an average error of 0.067%, and show that the proposed method achieves satisfactory performance,

https://doi.org/10.1515/9783112213094-007

with 86.34% accuracy, 82.83% regression rate, and 83.74% F-score. This approach successfully addresses the challenge of saving time and effort while accurately detecting the mass of liquid absorption over time, and provides a promising approach for a wider application of this intelligent model.

7.1.2 Methodology

Figure 7.1 illustrates the proposed framework for monitoring water absorption in composite structures. As shown in Figure 7.1, a high-quality solution addressing the drawbacks of present monitoring systems for estimating $M\%$ of glass fiber-reinforced epoxy laminated composite pipelines is proposed. This solution uses artificial intelligence algorithms for predicting $M\%$ using a dataset from finite element model analysis for the electrical capacitance sensor system to train a new deep neural network for different levels of $M\%$ not included in the finite element model data.

   The main contributions of this case study are outlined as follows:

1. A multi-physics finite element model is developed for electrical capacitance sensor electrodes installed around a glass fiber-reinforced epoxy laminated composite pipeline subjected to long-term internal pressure, to study the mechanical behavior over time (t).
2. Based on that finite element model, the $M\%$ of the composite pipeline material is analyzed for changes in the electric potential difference extracted between excited electrode pairs of electrical capacitance sensors at $t = 0$ days, $M\% = 0\%$, and for different levels of %.
3. A new deep neural network architecture is established and trained using changes in electric potential difference to predict various levels of $M\%$ not included in the finite element model datasets.
4. To verify the applicability and effectiveness of the present technique, a comparison between the results of the proposed approach, a dataset of experimental work established by d'Almeida et al. [10], and a theoretical model of a Fickian diffusion model presented by Ellyin and Maser [67] is utilized.
5. The results show an excellent agreement between the finite element model data, the available experimental and theoretical data, and the predictions made by the deep neural network, demonstrating that the proposed method achieves satisfactory performance.

7.1.2.1 The geometric model

The composite pipeline dimensions are considered to be 0.5 m, 0.04 m, and 0.043 m for length, internal radius (R_1), and outer radius (R_2) respectively. The stacking of fiber plies is specified, and the fiber layer thickness is 0.6 mm. The pipeline is subjected

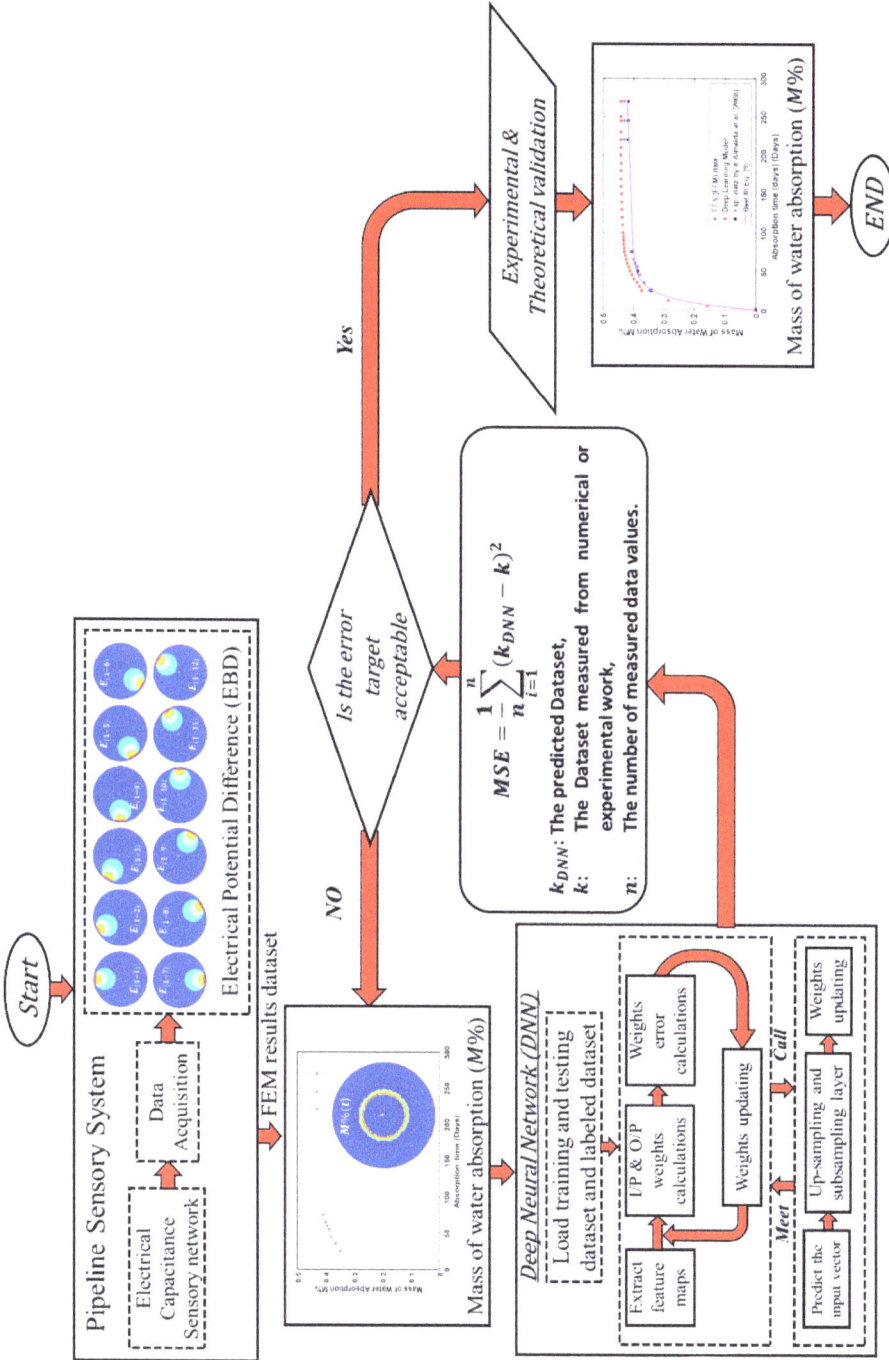

Figure 7.1: The block diagram of the proposed method.

to internal pressure, $P_0 = 1,500E4$ Pa. The liquid inside the composite pipeline has a temperature of 60 °C, where the outside temperature is 20 °C.

The basic principle for monitoring the mass of liquid absorption is that the capacitance of the sensor changes as the dielectric constant between the electrodes varies due to changes in the properties of the pipe material. For static excitation, the excitation voltage is 15 V. The measuring electrode remains 0 V, and the earth screen is connected to the ground.

Tables 7.1–7.3 show the properties of glass fiber-reinforced epoxy composite materials. These material properties are provided by a high-tech company, the National and Local Joint Engineering Research Center for Fiber Production and Application Technology, Nanjing, China. The company specializes in the research and development, manufacturing, marketing, and technical assessment of high-performance fibers and composites.

Table 7.1: Structural properties.

Element type	EX (Pa)	EY (Pa)	EZ (Pa)	PRXY	PRYZ	PRXZ	GXY (Pa)	GYZ (Pa)	GXZ (Pa)
SHELL181	53.74E9	17.95E9	17.95E9	0.25	0.5	0.25	5.98E9	3.78E9	8.63E9

Table 7.2: Thermal properties.

Element type	KX	KY	KZ	ALPX	ALPY	ALPZ	KXX	KYY	KZZ
		$(W{\cdot}m^{-1}{\cdot}K^{-1})$			(K^{-1})			$(W{\cdot}m^{-1}{\cdot}K^{-1})$	
SOLID87	0.18	0.08	0.08	8E-6	6E-6	6E-6	0.628	0.628	0.628

Table 7.3: Electrical properties.

Element type	Permittivity	PERX $(F{\cdot}m^{-1})$	PERY $(F{\cdot}m^{-1})$	PERZ $(F{\cdot}m^{-1})$	RSVX $(\Omega{\cdot}m)$	RSVY $(\Omega{\cdot}m)$	RSVZ $(\Omega{\cdot}m)$
SOLID123	Water	78.36	78.36	78.36	1E4	1E4	1E4
	Oil	3.0	3.0	3.0	3E11	3E11	3E11
	C_2H_5OH	24.5	24.5	24.5	7.4E6	7.4E6	7.4E6
	Glass fiber-reinforced epoxy	3.12	1.53	1.53	0.01	0.01	0.01
	Electrode	1E10	1E10	1E10	1.75E-6	1.75E-6	1.75E-6
	Air	1.0	1.0	1.0	3E13	3E13	3E13

Note: These electrical parameter properties can be found in references [411–413].

7.1.2.2 Numerical work

7.1.2.2.1 System assumptions

As stated in the main objective of this chapter, this case study presents the mass of liquid absorption monitoring in glass fiber-reinforced epoxy pipelines using electrical capacitance sensors. In industry, piping systems are generally exposed to high temperatures and high fluid pressure, and these loads are continuous and long term. Therefore, several assumptions are considered for numerical (finite element model) simulation and theoretical calculations to monitor the mass of liquid absorption in a glass fiber-reinforced epoxy pipeline. We can verify that those assumptions are fit for that purpose by comparing the finite element model and theoretical data with experimental data. The proposed assumptions are as follows:

1) The pipeline is filled with the inside fluid, and the pressure on the pipe wall is uniform and static.
2) The temperature of the pipe remains stable and does not affect the electrical capacitance sensors.
3) There are no localized small defects in the composite pipeline.
4) The pipeline has a thin wall (the thickness/Diameter of the pipeline is less than 0.1).
5) The electrode is assumed to be fully conductive, and the electrostatic field only affects the local area of the pipeline, one part of which is selected for analysis.

7.1.2.3 The structural–thermal–electrostatic (multi-physics) coupled field modeling

Figure 7.2 illustrates the finite element model of the glass fiber-reinforced epoxy laminate composite pipeline. The model is established based on three physical fields: structural, thermal, and electrostatic. Appropriate finite elements were selected and used to simulate glass fiber-reinforced epoxy properties, that is, we used SHELL 181, SOLID87, and SOLID123 element types to simulate the structural, thermal, and electrical properties, respectively, in ANSYS software. All the parameters required for the analysis of multi-physics coupled field are presented in Tables 7.1–7.3.

The multi-physics coupled field is established on the assumption that each physical field is created in a field with an independent fixed model and network. The coupled load is transferred by element surfaces or volumes, defining the problem and the order of the solution by a set of multi-field solver commands by automatically transferring the connected loads between different networks. In this study, the structural–thermal–electrostatic coupled field is established, and the mechanism of mutual load transfer is illustrated in Figure 7.3.

7.1.2.3.1 The static model analysis

As stated above, the internal pressure of the pipeline is $P_0 = 1,500E4$ Pa. The internal liquid temperature of the pipeline is 60 °C, and the external temperature is 20 °C.

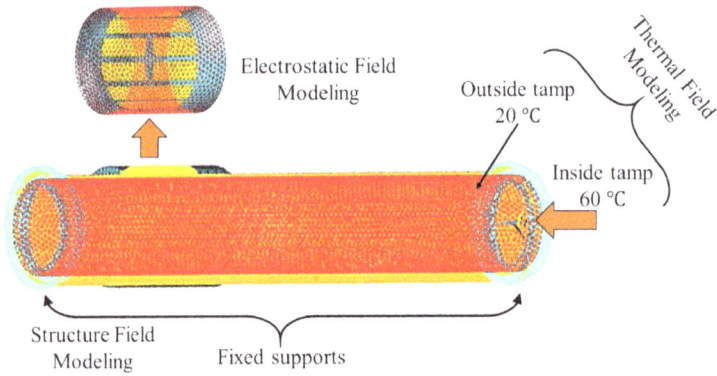

Figure 7.2: A finite element model of a glass fiber-reinforced epoxy laminate composite pipeline.

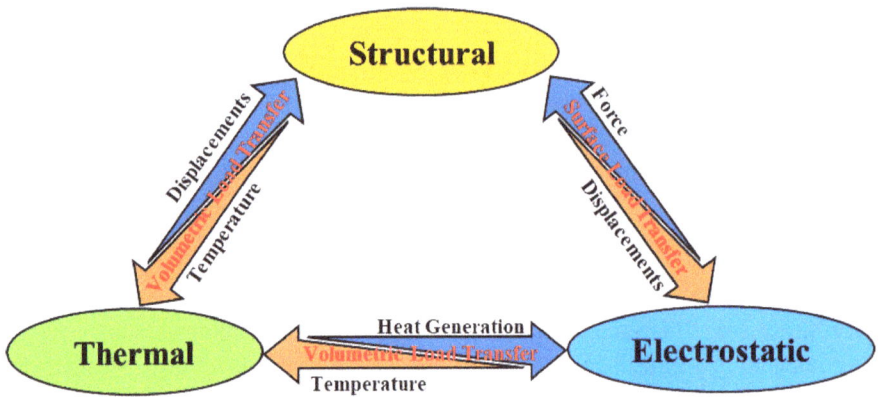

Figure 7.3: Schematic diagram of structural–thermal–electrostatic coupled field modeling.

Twelve electrical capacitance sensors are uniformly distributed around the external pipe wall. Under these specific conditions, the $M\%$ in the pipeline, and some basic analytical results regarding structural, thermal, and electrostatic coupled field distribution are presented as follows.

7.1.2.3.2 Electrostatic field results

The voltage distribution is presented in Figure 7.4. As shown in the figure, the maximum value of the electrostatic field intensity is observed at the exciting electrode of the electrical capacitance sensor network, and it diminishes as the distance from the exciting electrode increases.

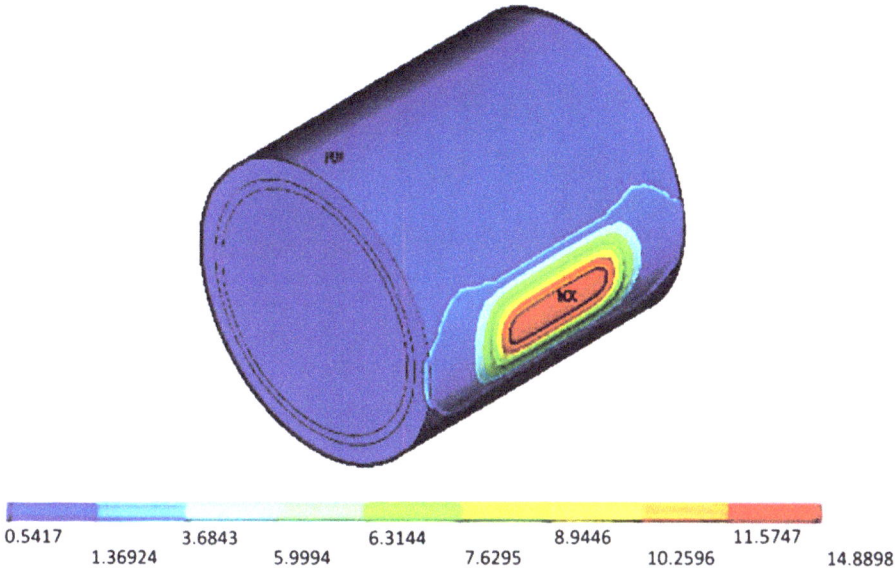

0.5417		3.6843		6.3144		8.9446		11.5747	

0.5417   3.6843   6.3144   8.9446   11.5747
       1.36924   5.9994   7.6295   10.2596   14.8898

Figure 7.4: The potential distribution in volts.

7.1.2.4 The mass of water absorption (*M%*) monitoring

From the analysis results, using the scripting capabilities in ANSYS, we apply the basic functionalities of the Fickian diffusion model to describe the water absorption data in Matlab code that will help us in developing scripts in ANSYS Discovery for the simulation of *M%* levels. We plot these results using finite element model data analysis at any absorption time (*t*) before saturation for the detected amount of *M%* by accessing different objects in ANSYS Discovery from the number of elements at any absorption time (*t*) before saturation (E_t) and the number of elements fabricated (E_F) ($t = 0$ days) using the formula:

$$M\% = \frac{E_t - E_F}{E_F} \times 100 \qquad (7.1)$$

For example, Figure 7.5 shows *M%* detected by finite element model data analysis at 50 days. From Figure 7.5, and by applying Equation (7.1), we can calculate $M\% = 0.38\%$ at 50 days.

The simulation model was performed with a fabricated (time of absorption, $t = 0$ in days) and an aged (saturation was reached) glass fiber reinforced epoxy pipes. The absorption times for the aged pipes were selected based on the water absorption characteristics of the composite, with constant working conditions and water pressure of $P_0 = 1{,}500E4$ Pa. The internal liquid temperature of the pipeline was 60 °C, and the external temperature was 20 °C. Figure 7.6 shows the effect of the mass of water absorption (*M%*) at absorption time variation (*t* in days) on the node potential distribution of the

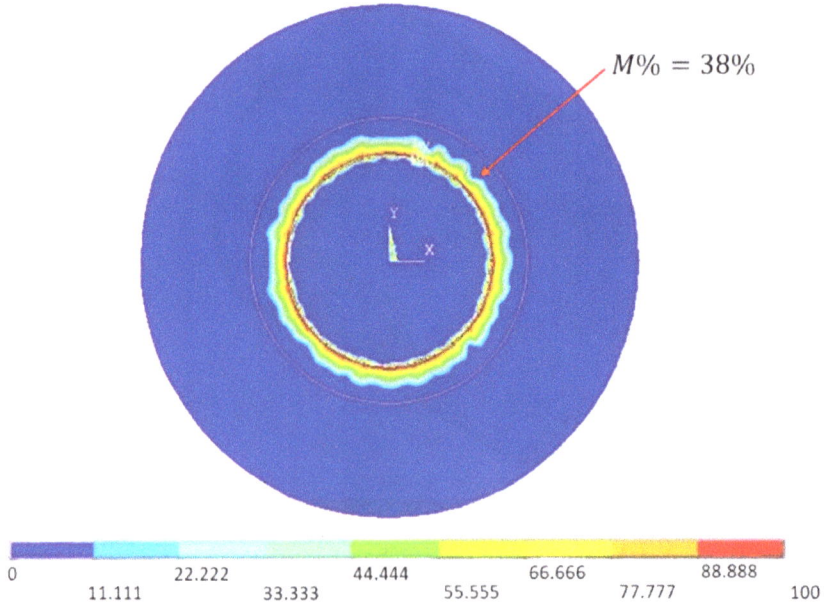

Figure 7.5: Mass of water absorption (*M*%) monitoring from finite element model at 50 days.

first electrode for three stages of the water absorption rate in a pressurized glass fiber-reinforced epoxy composite pipe: the as-fabricated (days, $M\% = 0\%$) at $t = 0$ days, during absorption stage ($M\% = 56.8\%$) at $t = 75$ days, and the saturation stage ($M\% = 99.6\%$) at $t = 245$ days, respectively.

There are two areas of node potential distributions across the pipe: the blue areas or $\varphi = 0$ (regions without node potential) and the colored areas or $\varphi > 0$, $\varphi \leq V_0$ (regions with different node potentials), that is, the colored areas represent the electrode-sensitive domain (detection domain).

Based on the analysis of Figure 7.7, we can conclude that there is a noticeable difference in node potential distributions and electric field intensity in three different stages of water absorption rate with absorption times.

As water absorption progresses through the glass fiber-reinforced epoxy composite pipe, the corresponding capacitance of the electrode pair keeps a relatively stable value (see curves in Figure 7.9). From Figure 7.9, we can observe that the pipe material began to be saturated after 75 days. However, when the absorption times were between 25 and 75 days, saturation was not yet reached (i.e., the saturated pipe material was observed for the longer absorption times).

(a) Fabricate, $M\% = 0\%$ at $t = 0$ days

(b) During absorption, $M\% = 56.8\%$ at $t = 75$ days

(c) Saturation stage, $M\% = 99.6\%$ at $t = 245$ days

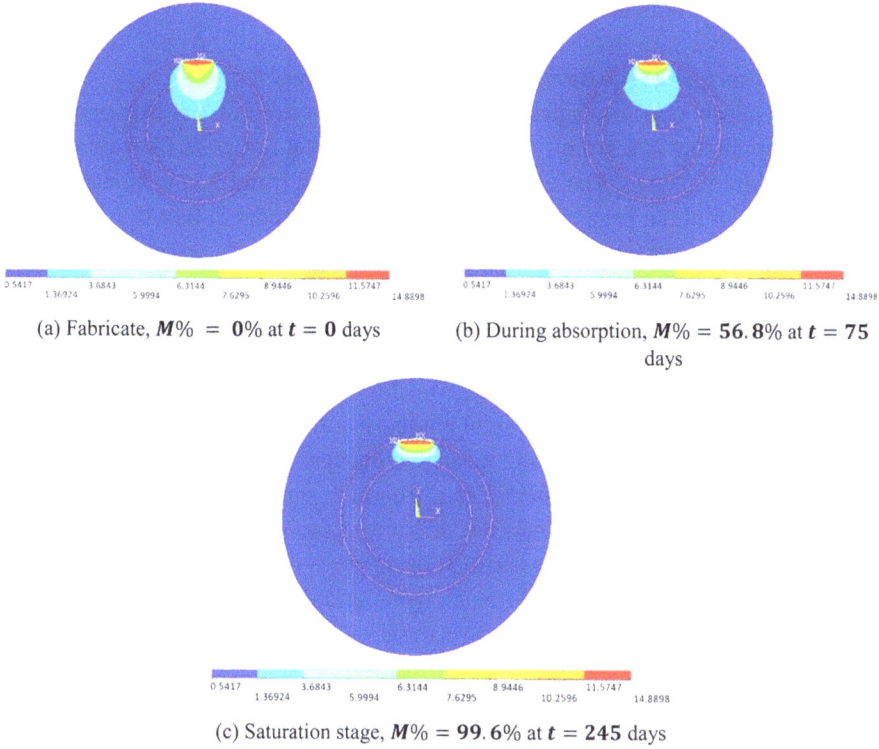

Figure 7.6: The node potential distribution of the pipe material in volts.

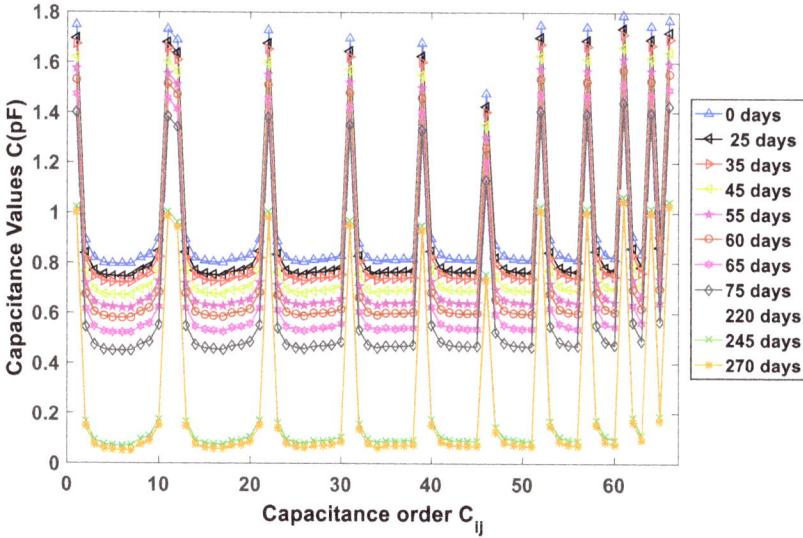

Figure 7.7: The corresponding capacitance of the electrode as a function of absorption time (days).

7.1.2.5 Deep neural networks

7.1.2.5.1 Architecture of the developed deep neural network

A multilayer neural network is developed to construct a deep neural network model capable of predicting $M\%$ from finite element model data of glass fiber-reinforced epoxy composite pipelines. The network is designed and trained in systematic form for generalization of the proposed deep neural network model to find the promising deep neural network architecture.

The proposed deep neural network architecture with sequential hidden layers can be described, as shown in Figure 7.8. Fully connected layers are located between the input (features) and output layers (target prediction) for higher-level feature extraction via the process of training. Using weights W and biases b, the deep neural network is trained to identify more useful nonlinear information.

One-dimensional convolution and mean pooling operations are utilized to extract features from the finite element model data. The fully connected neural network includes 6 hidden layers, with the number of neurons for the first 3 layers being 502 and the other three layers 251. The rectified linear units (ReLUs) are used as the activation function for both the convolutional layers and the fully connected layers. In the proposed deep neural network, this process is defined as

$$x_j^l = f\left(\sum_i x_i^{l-1} w_{ij}^l + b_j^l \right) \tag{7.2}$$

where x_j^l is the ith output map in layer l; x_i^{l-1} is the ith output map in layer $l-1$; w_{ij}^l is the weight; b_j^l is the bias; $f(\cdot)$ is a nonlinear function that is applied component-wise.

Figure 7.9 shows the basic architecture of the convolutional layers used in this study, including the subsampling layer. $C1$ is a convolution layer that comprises six feature maps.

Table 7.4 shows the convolution layer's detailed settings, where the first convolutional layer, $C1$, consists of six feature maps. The connection size of each feature map neuron is a 9×9 input matrix. The size of the feature maps is 99×99 matrices. There are four filters in total, and each filter has 9×9 unit parameter weights and a bias parameter, totaling $(9 \times 9 + 1) \times 4 = 328$ tuned parameters. Between the input and the first convolutional layer, $C1$, one kernel is used; therefore, the total turns out to have $328 \times (99 \times 99) = 3{,}214{,}728$ connections. A subsampling layer $S1$ has $(1+1) \times 2 = 4$ tuned parameters and connections, and so on for the next two layers, $C2$ and $S2$, which have similar architectures, as shown in Table 7.4.

7.1.2.5.2 Deep neural network training and test sets

A finite element model result data by ANSYS in Section 7.1.3.3.2 is used to obtain the training data and predict the water $M\%$ levels of a glass fiber-reinforced epoxy

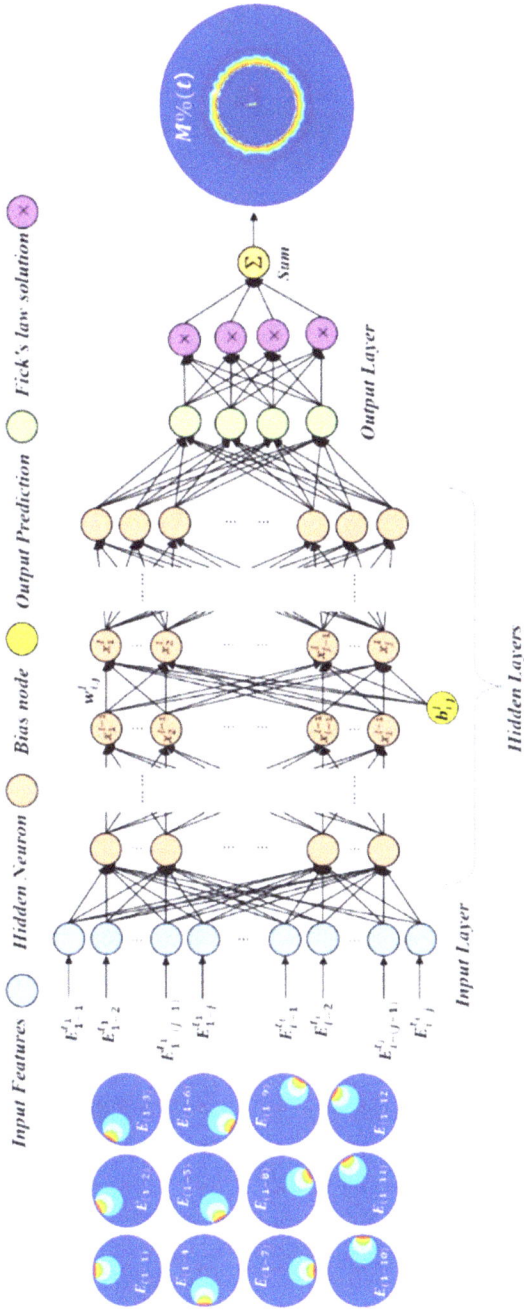

Figure 7.8: The architecture of the proposed deep neural network to predict M% with input, sequential hidden, and output layers.

Table 7.4: Detailed settings of the convolutional layers.

Layer	Kernel size	Number of parameters	Number of connections
C1	9×9	148	340,992
S1	4×9	4	34,816
C2	9×9	//	//
S2	4×4	//	//

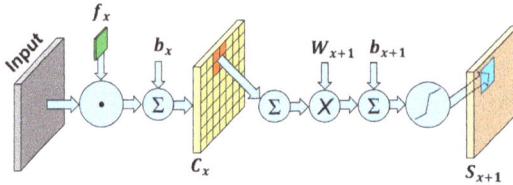

Figure 7.9: Connection of convolutional layer and subsampling layer.

composite pipe under internal hydrostatic pressure load as a function of absorption time (t) (see Figure 7.1).

Figure 7.10 shows the proposed algorithm that works with adaptive features and variable parameters to operate in both training and testing modes. The training mode starts from the results data of finite element model to adjust the parameters that control the hidden layers. Figure 7.10 consists of two stages: training with labeled data and testing with known labeled data (expected) in the right and the left, respectively. The number of hidden layers can increase or decrease based on the results of the training mode. The actual data comes after processing the results several times, as iterations of testing through deep neural network until satisfactory results are obtained, and these are discussed in the next section. The deep neural network setup, training, and testing model steps are presented in Figure 7.11.

The developed deep neural network model is fed with 36,877 samples of training data resulting from the finite element model of the electrical capacitance sensor analysis. The preparation of the training samples is described in detail as follows: the number of samples fabricated is 21,642 ($M\% = 0$) and the number of samples during water absorption is 15,235 ($M\% \neq 0$), 80% are randomly selected as the training set, and the remaining 20% serves as the test set. Subsequently, the pre-training network is used for migration learning, all data are rotated 500 epochs, and the mean squared error is utilized as the loss function (see Figure 7.1). Figure 7.12 presents the training and test mean squared error losses under supervised mode. It can be observed that the mean squared error drops to small values with the increase in the training epoch and the losses associated with $M\%$ exhibit small values and decrease slowly after 10 epochs. In addition, the test mean squared error shows larger values compared to the

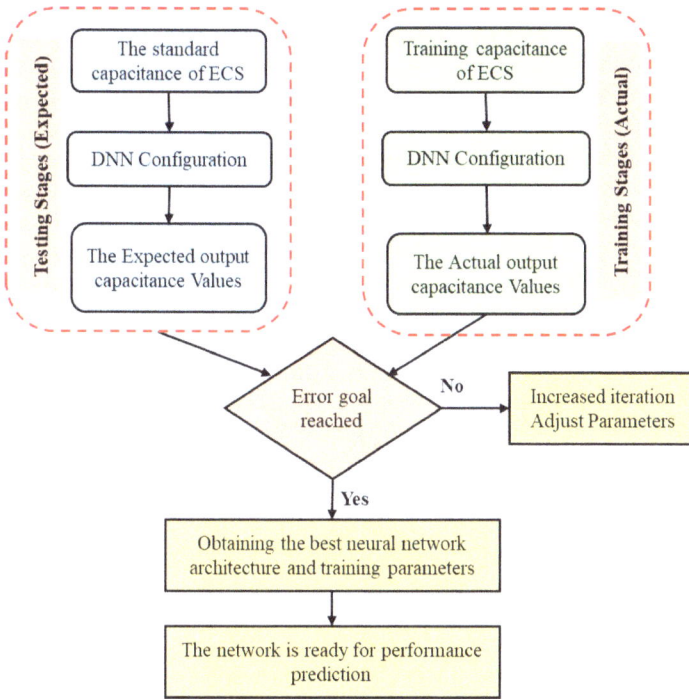

Figure 7.10: Flowchart of the proposed method.

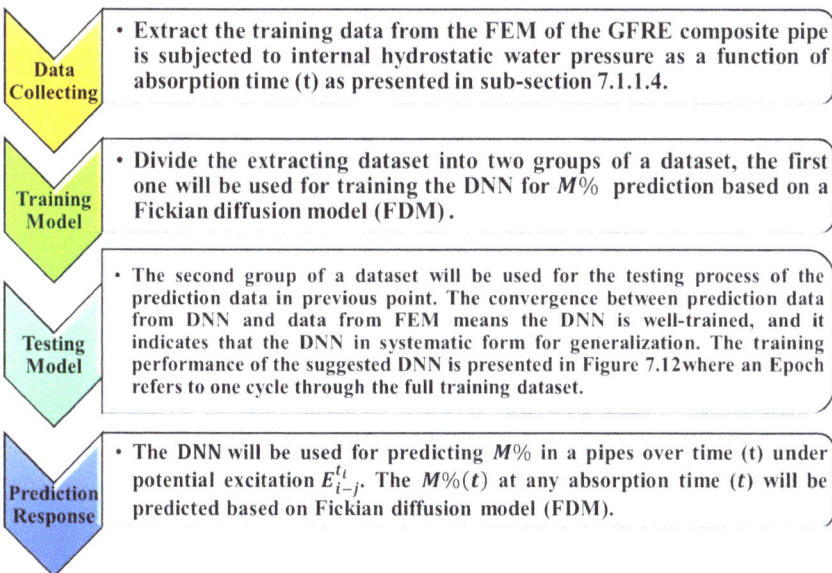

Data Collecting	• **Extract the training data from the FEM of the GFRE composite pipe is subjected to internal hydrostatic water pressure as a function of absorption time (t) as presented in sub-section 7.1.1.4.**
Training Model	• **Divide the extracting dataset into two groups of a dataset, the first one will be used for training the DNN for $M\%$ prediction based on a Fickian diffusion model (FDM).**
Testing Model	• **The second group of a dataset will be used for the testing process of the prediction data in previous point. The convergence between prediction data from DNN and data from FEM means the DNN is well-trained, and it indicates that the DNN in systematic form for generalization. The training performance of the suggested DNN is presented in Figure 7.12 where an Epoch refers to one cycle through the full training dataset.**
Prediction Response	• **The DNN will be used for predicting $M\%$ in a pipes over time (t) under potential excitation $E_{i-j}^{t_i}$. The $M\%(t)$ at any absorption time (t) will be predicted based on Fickian diffusion model (FDM).**

Figure 7.11: The steps of deep neural network training to predict $M\%$.

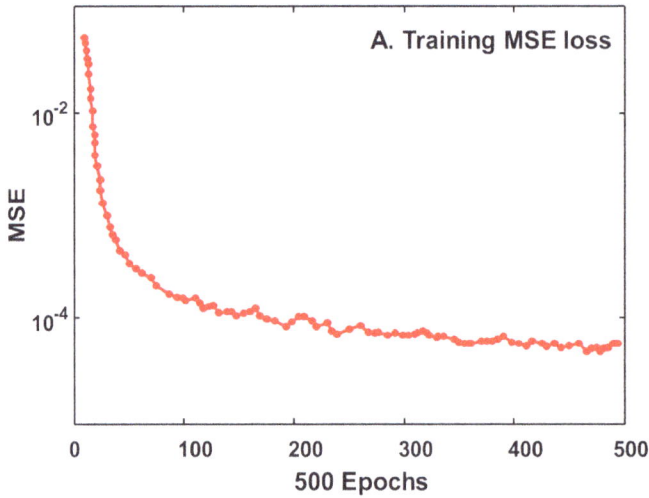

(a)

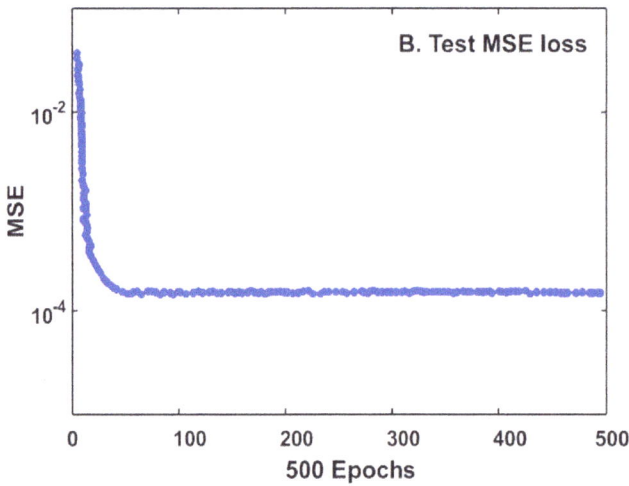

(b)

Figure 7.12: Training and testing mean squared error loss of suggested deep neural network.

training mean squared error, while it can reach a stable state earlier. Overall, better training and test performance of the developed deep learning model can be observed for the $M\%$ of glass fiber-reinforced epoxy composite pipe. The key parameters for deep neural network training are shown in Table 7.5.

Table 7.5: Key parameters for deep neural network training.

Training time	Training time	Training rate	Attenuation factor
35 s	2.25 s	10^{-4}	10^{-6}

7.1.2.5.3 The training and testing algorithm

The establishment, training, and testing of a deep neural network were carried out using MATLAB software, and the neural network was developed following the steps outlined in Algorithm 7.1.

Algorithm 7.1: Training and testing of the proposed deep neural network.

1: **input**: d: $M\%$ dataset, I: permittivity ε, EPD, W: Network parameter matrix weight w_{ij} and bias b_j
2: **output**: score of deep neural network trained model on the dataset to predict $M\%$ for various ε, EPD
3: **let** f be the feature set 3d matrix
4: **for** i in the dataset **do**
5: **let** f_i be the feature set matrix of sample l
6: **for** j in i **do**
7:   $V_i \leftarrow vectorize_{(j,w)}$
8:   **append** V_i to f_i
9:   **append** f_i to f
10: $f_{train}, f_{test}, l_{train}, l_{test} \leftarrow$ the split feature set and prediction into train subset and test subset.
11: $M \leftarrow$ deep neural network (f_{train}, l_{train})
12: score $\leftarrow$ evaluation (l, l_{test}, M)
13: **return** score
14: **end for**
15: **end for**

7.1.2.5.4 The developed deep neural network predicted data of $M\%$

Figure 7.13 shows a comparison between the mass of water absorption $M\%$ curve predicted from the proposed deep neural network and the results by ANSYS.

As illustrated in the figure, $M\%$ curve obtained from the deep learning model shows good convergence with the results from the finite element model. Therefore, $M\%$ of composite pipes can be predicted by the deep neural network model studied in this case study. This validates the accuracy and reliability of the proposed technique. From Figure 7.13, according to the deep learning model, we can observe that the absorption time from 5 days to 80 days had not yet reached saturation. However, the pipe material began to saturate after approximately 80 days, and the mass of water absorption reached its saturation level at $M\% = 43\%$ at $t = 270$ days.

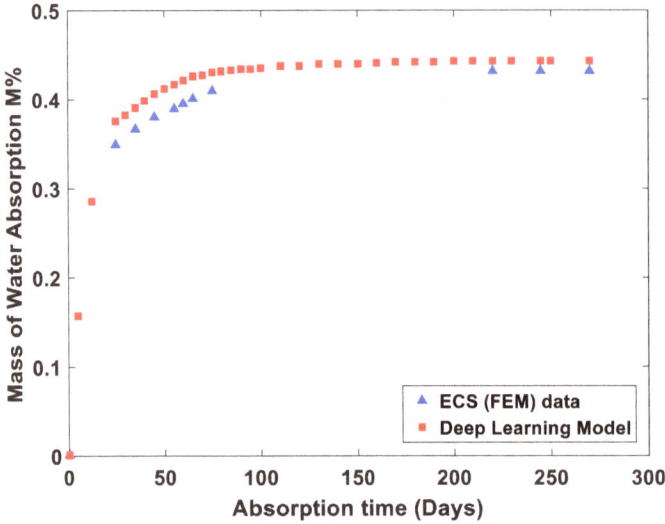

Figure 7.13: Mass of water absorption (*M*%) as a function of absorption time (days).

7.1.2.6 Validity of the proposed technique

7.1.2.6.1 Theoretical validation

A Fickian diffusion model can describe the water absorption data [414], as shown in Figure 7.16. Fick's law can be applied under uniform initial conditions, assuming that the pipe geometry exhibits unidirectional diffusion through its thickness (e.g., a thin pipe $(h/d) \leq 0.1$). Fick's law is then expressed as described by Springer [415]:

$$\frac{M\%}{M_\infty} = 1 - \frac{8}{\pi^2} \sum_{i=0}^{\infty} \frac{1}{(2i+1)^2} \exp\left[-D(2i+1)^2 \pi^2 \frac{t}{h^2}\right], \quad h = (R_2 - R_1) \tag{7.3}$$

and

$$D = \pi \left(\frac{kh}{4M_\infty}\right)^2, \quad k = \left(\frac{M_2 - M_1}{\sqrt{t_2} - \sqrt{t_1}}\right)^2 \tag{7.4}$$

For $(M\%/M_\infty) < 0.5$, the complete absorption curve of Fick's law can be represented by McKague et al. [416] using the simplified equation from Equation (7.3):

$$\frac{M\%}{M_\infty} = \tanh\left(\frac{4}{h}\sqrt{\frac{Dt}{\pi}}\right) \tag{7.5}$$

7.1.2.6.2 Experimental validation

In this section, we apply the proposed approach to a dataset of experimental work adapted from d'Almeida et al. [417] for glass fiber composite pipes used in service waters. These pipes are directly exposed to water for long periods and must maintain their integrity without losing their mechanical performance.

Figure 7.14 shows a comparison of the mass of water absorption $M\%$ between the electrical capacitance sensor results from the finite element model, the proposed deep neural network predicted data, the experimental data by d'Almeida et al. [417], and the theoretical data from the Fickian diffusion model in Equation (7.5) for the same pipe's geometric specifications, initial conditions, and absorption times. Figure 7.15 presents the correlation between the experimental results and the deep neural network predictions, demonstrating excellent agreement between the outcomes of the proposed method and the experimental data, with an average error of 0.067%. As shown in Figures 7.14 and 7.15 the results exhibit good convergence among the finite element model, d'Almeida's experimental data, the Fickian diffusion model, and the deep neural network predictions. Furthermore, the findings indicate that the proposed method can successfully monitor the mass of liquid absorption $M\%$ in laminated composite pipelines.

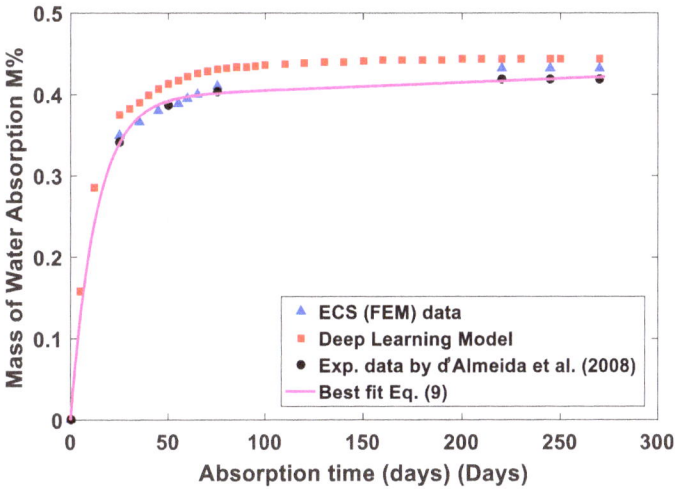

Figure 7.14: The mass of water absorption ($M\%$) as a function of absorption time (days).

7.1.2.6.3 Evaluation of present algorithm accuracy and reliability

To estimate the performance of the proposed method for monitoring the mass of water absorption in laminated composite pipes, three indicators are calculated during the training process: the accuracy rate ($P\%$), the regression rate ($R\%$), and the F-score ($F\%$). The accuracy rate is defined as the percentage of correctly classified data

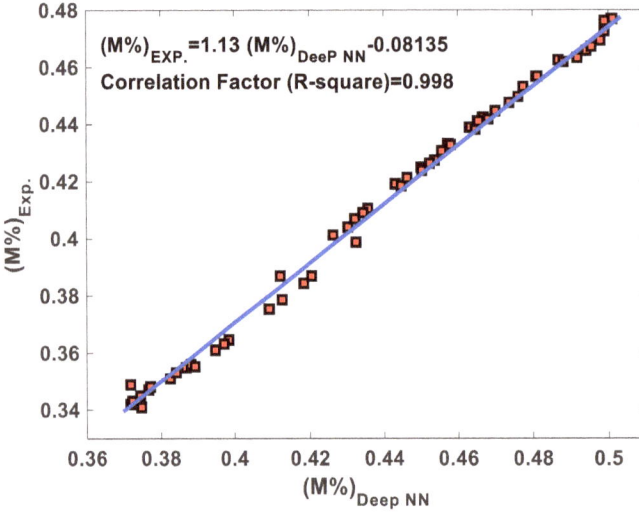

Figure 7.15: Correlation of experimental and deep learning results.

among all monitored data in $M\%$ results of electrical capacitance sensors; the regression rate is defined as the percentage of correctly classified infected data out of all actual infected data in $M\%$ results of electrical capacitance sensors. The F-score is defined as a measure of the training accuracy [418, 419].

To check the present deep neural network accuracy and its effectiveness for monitoring the mass of water absorption ($M\%$) in composite pipes, big data on the corresponding capacitance of the electrical capacitance sensor electrodes were utilized. Thirty-three percent of this data was selected as samples for computing $P\%$, $R\%$, and $F\%$. Ten datasets of $M\%$ were established from the testing samples, denoted as $S1$, $S2$, ..., and $S10$. To compute $P\%$, $R\%$, and $F\%$, four indices, including the true positive rate (TPR), true negative rate (TNR), false positive rate (FPR), and false negative rate (FNR), were determined:

$$P\% = \frac{N_{\text{TPR}}}{N_{\text{TPR}} + N_{\text{FPR}}} \tag{7.6}$$

$$R\% = \frac{N_{\text{TPR}}}{N_{\text{TPR}} + N_{\text{FNR}}} \tag{7.7}$$

$$F\% = \frac{2N_{\text{TPR}}}{2N_{\text{TPR}} + N_{\text{FNR}} + N_{\text{FPR}}} \tag{7.8}$$

Figure 7.16 and Table 7.6 present the resultant computed $P\%$, $R\%$, and $F\%$ versus the number of iterations. The 10 dataset results of $M\%$ (i.e., $S1$, $S2$, ..., and $S10$) are indicated with solid lines with different markers, whereas the overall performances of selected $M\%$ are indicated with dotted lines. As we can see from the plots in Figure 7.18, the highest and lowest orders of $P\%$ are denoted for $S9$ and $S10$, respectively, and the highest and lowest orders of R are denoted for $S9$ and $S6$, respectively. This means that they have the highest and lowest values of $M\%$, and the other groups of $P\%$ and $R\%$ fall between these values.

Table 7.6 lists the average values of $P\%$, $R\%$, and $F\%$ for each dataset's results of $M\%$ (i.e., $S1$, $S2$, . . ., and $S10$). The values in this table are derived from Figure 7.16, where the averages of the dataset results are indicated by the black dotted line. In general, $P\%$, $R\%$, and $F\%$ versus the overall performance are 86.34%, 82.83%, and 83.74%, respectively. The results confirm that the proposed method can automatically monitor the mass of water absorption in laminated composite pipes with satisfactory performance, regardless of the corresponding capacitance data noise backgrounds and conditions.

Table 7.6: Comprehensive evaluation results for the automatic mass of water absorption monitoring algorithm.

Datasets	P (%)	R (%)	F (%)
S1	84.75	82.05	81.856
S2	89.27	82.45	84.19
S3	80.91	76.84	78.49
S4	87.96	85.16	84.86
S5	83.86	84.21	83.65
	84.47	75.69	79.49
S7	90.36	82.52	85.87
S8	91.37	80.55	85.24
S9	93.56	94.73	93.75
S10	76.91	84.07	79.99
AVG	**86.34**	**82.83**	**83.74**

7.1.2.7 Summary

A deep learning model is proposed for predicting the mass of water absorption ($M\%$) in glass fiber-reinforced epoxy composite pipes under internal hydrostatic pressure load and the thermal effect of water over absorption time (t). A new deep neural network, established with a fully connected sequential hidden layer, is positioned between the input (features) and output layers (target prediction) for predicting $M\%$ based on the Fickian diffusion model as a function of absorption time (t). The effect of $M\%$ over time (t) based on the capacitance signals generated between the electrical capacitance sensor electrode pairs due to dielectric property changes in the pipeline material was evaluated using a finite element model analysis. The training and testing datasets for the suggested deep neural network were generated by the finite element model analysis of the installed electrical capacitance sensors around the outer diameter of the composite pipe. The training performance of the new deep neural network was evaluated using four indices: TPR, TNR, FPR, and FNR. The results of the deep

(a) Accuracy

(b) Regression rate

(c) F-score

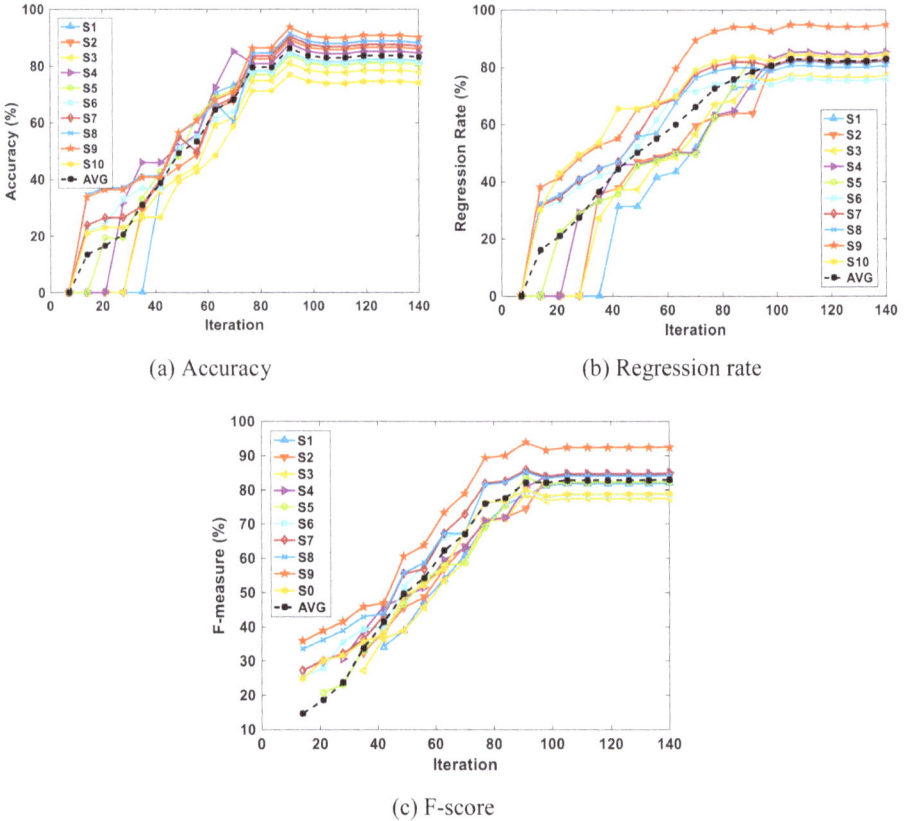

Figure 7.16: Comparison of the training process based on the mass of water absorption monitoring test data for 10 datasets of *M*%.

learning model for of *M*% were verified using datasets from experimental work adapted from the literature and the finite element model of the electrical capacitance sensor system. The following conclusions can be drawn from the results:

1. In this case study, it is proved that the proposed deep learning model can predict the *M*% of the composite pipeline, as shown in Figures 7.14 and 7.15.

2. The main drawback of numerical and experimental methods in the analysis of electrical capacitance sensor datasets has been overcome, where the generated electric potential difference between the electrode pairs of electrical capacitance sensors depends on the changes in *M*% within the pipeline material, which takes a long time to reach the saturation point. The proposed method works much faster than both experimental methods and finite element models in terms of data processing and computational time. Therefore, the suggested method has certain advantages to the electrical capacitance sensor technique, for improving *M*% monitoring in composite structures.

3. The results of the suggested deep neural network under structural, thermal, and electrostatic parameters of pipeline presented in this case study show that the saturation point of the pipeline material (glass fiber-reinforced epoxy) with water begins after 80 days. However, during the period between 5 days and 80 days, the mass of water absorption inside the pipeline material increases from 17% to saturated final point that occurs at $M\% = 43\%$, $t = 270$ days.
4. The results of the suggested deep neural network show a good agreement with the experimental dataset adapted from the literature and the finite element model dataset of the electrical capacitance sensor system with an average error of 0.067%, and with an acceptable correlation factor (CF) between the deep neural network and the experimental dataset is 0.998, which is very close to unity.

    The values of four indices that were used to evaluate the accuracy rate ($P\%$), regression rate ($R\%$), and F-score ($F\%$) of the deep neural network training process indicate satisfactory performance, with high values for $P\%$, $R\%$, and $F\%$ as 86.34%, 82.83%, and 83.74%, respectively. This validates the accuracy and reliability of the proposed deep neural network.

7.1.3 Case study (2): predicting long-term creep thermomechanical fatigue behavior monitoring in composite pipelines using electrical capacitance sensors integrated with deep learning approach

The composite pipeline is a relatively new and viable alternative pipeline to the more commonly used traditional one due to its good mechanical and fatigue properties, and lower production cost. For this reason, it is critical to assess the mechanical and fatigue performance of composite pipeline materials under various working conditions, particularly for monitoring long-term creep thermomechanical fatigue (LTCTMF) behavior. In this case study, the LTCTMF behavior of a basalt fiber-reinforced polymer laminated composite pipeline is detected through an integrated expert system consisting of electrical capacitance sensors and a deep learning algorithm. First, a multi-physics finite element model is developed to simulate the LTCTMF behavior of basalt fiber-reinforced polymer composite pipelines subjected to prolonged fatigue loading caused by internal pressure and thermal effects. Second, theoretical model results of LTCTMF compliance $(S_f(t))$ over the time of creep are analyzed in the pipeline material using the modulus degradation approach. Finally, an electrical potential change between electrical capacitance sensor electrodes corresponding to $S_f(t)$ over the time of creep for some levels of LTCTMF ($R_f\%$) is recorded, and then used in these datasets for training the novel deep neural network based on one of the most widely used deep neural network families called convolutional neural network (CNN), to predict $S_f(t)$ in pipelines for various $R_f\%$ not included in the previous finite element model evaluation (i.e., electrical capacitance sensor technique). In this case study, the LTCTMF behavior is first detected for $R_f\% = 25\%$, 50%, and 75% from a finite element model and a modulus degradation approach, and then the LTCTMF behavior for $R_f\% = 15\%$, 35%, 60%, and 85%, respectively, via deep neural network is predicted. The results of the proposed method are in good agreement with experimental results available in the literature, thereby verifying the accuracy and reliability of the proposed technique and its applicability to other different composite structures.

7.1.3.1 Methodology

From an open scientific literature survey and to the best of the authors' knowledge, no work has been conducted related to a new LTCTMF indicator based on electrical capacitance sensors for composite pipeline materials. Therefore, the main goal of the current research is to propose a new framework integrating the electrical capacitance sensor technique with deep learning schemes for monitoring the LTCTMF behavior over creep time in basalt fiber-reinforced polymer laminated composite pipelines under various $R_f\%$ of fatigue stress (S_f) as a percentage of fatigue strength (U_f). Figure 7.17 shows the proposed framework for LTCTMF behavior monitoring in basalt fiber-reinforced polymer composite pipelines. As shown in Figure 7.17, the electric potential difference between electrical capacitance sensor electrodes corresponding to $S_f(t)$ over creep time is measured for three different $R_f\%$ to provide the basic dataset for training a novel deep neural network based on a CNN architecture. Subsequently, the trained deep neural network is utilized to predict $S_f(t)$ over creep time for other various $R_f\%$ not included in the electrical capacitance sensor measurement data. The results of the proposed method are verified theoretically using the modulus degradation approach to simulate LTCTMF in pipeline materials and experimentally using the experimental results available in the literature. The results show an excellent agreement between the proposed method's results and the predicted, theoretical, and experimental results.

7.1.3.2 The geometric model

The composite pipe specimens in this research have 0.5 m, 0.04 m, and 0.043 m for length, internal radius of (R_1), and an outer radius (R_2), respectively, the specimens consist of five layers, each with 0.6 mm thickness for each, and the fiber plies stacking is $[-45, 45, -45, 45, -45]_s$. The LTCTMF tests are conducted on a basalt fiber-reinforced polymer laminated composite pipeline subjected to long-term fatigue loading due to internal pressure and the thermal effect caused by the temperature difference between the inside and outside pipe wall temperature, whereas the internal pressure is $P_0 = 1,500E4$ Pa, and the liquid inside the pipe has a temperature of 60 °C, while the outside temperature is 20 °C.

Tables 7.1–7.3 introduce the structural–thermal–electrical properties of basalt fiber-reinforced polymer composites, and the material characteristics are provided by the National and Local Joint Engineering Research Center for Fiber Production and Application Technology, Nanjing, China. The center is engaged in the research

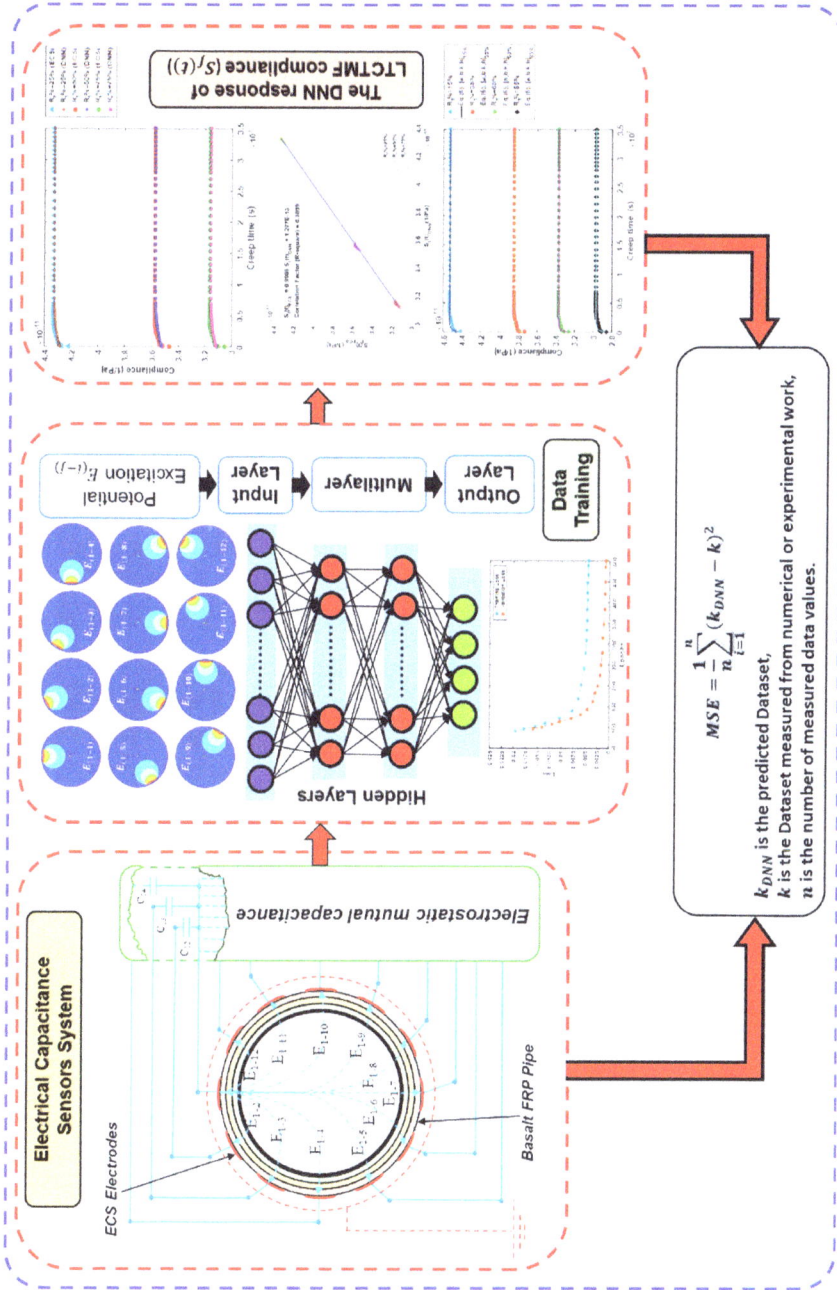

Figure 7.17: The block diagram of the proposed method for monitoring LTCTMF behavior in the basalt fiber-reinforced polymer composite pipeline.

and development, manufacturing, marketing, and technical evaluation of high-performance fibers and composite materials.

7.1.3.3 Numerical work

Figure 7.2 shows the finite element model of the basalt fiber-reinforced polymer laminate composite pipeline, and the finite element network was selected to fit each multi-physics field used in this case study, that is, the element types selected for the structural, thermal, and electrical fields are SHELL 181, SOLID87, and SOLID123, respectively. The finite element model was developed using ANSYS software.

Figure 7.3 presents a finite element model framework used in this case study. It is a combined structural–thermal–electrostatic field, and it is called multi-physics finite element model. As shown in Figure 7.3, each field in multi-physics has an independent fixed network and model. The figure shows the load transfer mechanism between each other by element surfaces, such as the force between the structural and electrostatic fields or the displacement for opposite transfer, and by automatically transferring connection loads between different networks, problems, and solution sequences detected through a set of multi-field solver commands.

7.1.3.3.1 The structural–thermal–electrostatic modeling analysis

As shown in Table 5.2, which summarizes various failure criteria, it was found that the von Mises stress and strain criteria are the most commonly used for studying fatigue failure in composite materials [420, 421]. In this case study, the von Mises stress and strain criteria were selected to study the fatigue failure in a basalt fiber-reinforced polymer laminated composite pipeline subjected to long-term fatigue loading of internal pressure and the thermal effects of the pipe temperature difference between the inner and outer pipe walls. In this section, the results of the structural–thermal–electrostatic finite element model analysis, as shown in Figure 7.18, indicate that the maximum equivalent stress (von Mises stress) and displacement occur near the fixed end supports. This suggests more severe damage in that region, where the maximum displacement value is 0.121×10^{-3} m and the maximum stress is 0.294×10^3 MPa.

Figure 7.19 presents the thermal and voltage field distributions. As shown in Figure 7.19(a), there is a uniform distribution of heat across the pipe thickness, where the maximum temperature at the inner pipe wall equals 60 °C and the minimum temperature at the outer pipe wall equals 20 °C. The voltage distribution around the excitatory electrode of the electrical capacitance sensors shows that the voltage intensity increases as we move closer to the excitatory electrode and diminishes as the distance from the excitatory electrode increases, as shown in Figure 7.19(b).

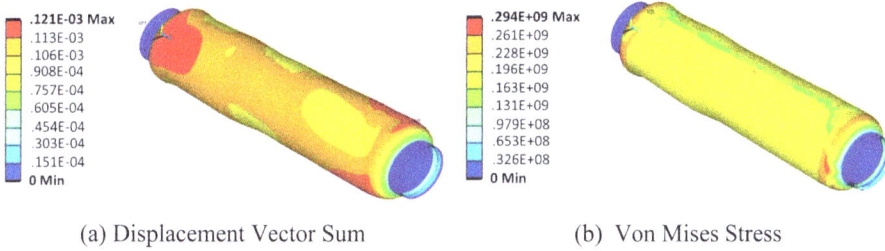

(a) Displacement Vector Sum

(b) Von Mises Stress

Figure 7.18: The finite element model analysis of the structural field.

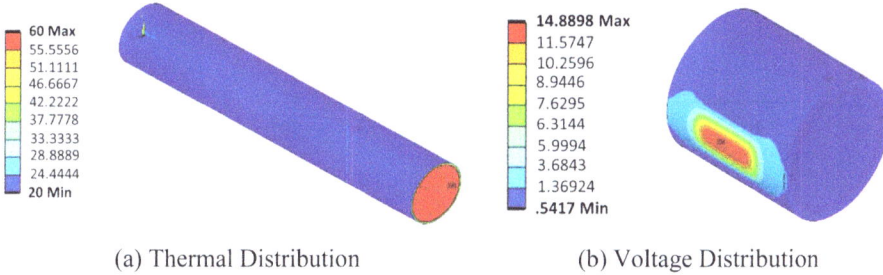

(a) Thermal Distribution

(b) Voltage Distribution

Figure 7.19: The finite element model analysis of thermal and electrostatic effects.

7.1.3.3.2 *S–N* curve of fatigue behavior for basalt fiber-reinforced polymer composite pipeline

The basalt fiber-reinforced polymer composite pipeline is subjected to long-term fatigue loading due to pressure (P) and the thermal effects caused by the temperature difference ($\Delta\theta$) between the inside and outside pipe wall. The simulation of fatigue models in the pipeline was performed using ANSYS for different pressure ratios (P_r) between the applied (P) and the burst pressure (P_{max}) which is equal to (0.5, 0.7, 0.9). The *S–N* curve is plotted between the number of cycles to failure N_f versus fatigue ratio (R_f) between the fatigue stress (σ_f) and the fatigue strength (S_f).

By using a built-in MATLAB curve-fitting routine to fit the *S–N* curve data using the power formula $R_f = aN_f^b$, it found that it was suitable by giving a high CF near unity equal to 0.9731, 0.9313, 0.9002 for $P_r = 0.5$, 0.7, 0.9, respectively. The constants of that formula, known as the fatigue constants, are listed in Figure 7.20 of the *S–N* curve for each P_r.

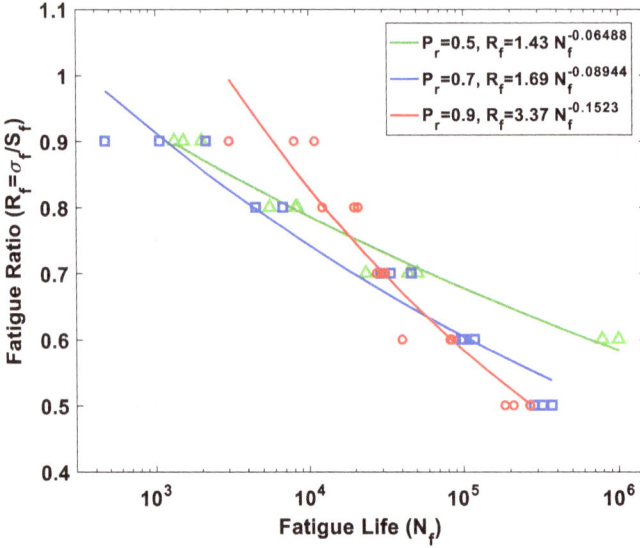

Figure 7.20: S–N curve of basalt fiber-reinforced polymer composite pipeline for various P_r.

7.1.3.3.3 D–N curve of fatigue damage model for basalt fiber-reinforced polymer composite pipeline

The D–N curve for the cumulative damage index of composite laminates in relation to relevant factors can be calculated using the following formula:

$$D = \left\{ \left(1 - \frac{F}{E_c(T)}\right)(1 - f^*)\frac{ln(N+1)}{ln(nN_f)} \right\} + \left\{ \left(1 - \frac{F}{E_c(T)}\right)f^*\left(\frac{N}{nN_f}\right) \right\}$$

$$+ \left\{ \frac{F}{E_c(T)}\left(1 - \frac{\sigma_{max}(1-R)}{2\sigma_{ult}(T)}\right)\frac{ln\left(1 - \frac{N}{nN_f}\right)}{ln\left(\frac{1}{nN_f}\right)} \right\} \qquad (7.9)$$

where D is the damage index, $N_f(T)$ is the fatigue life, f^* is the fiber matrix interface strength, $\sigma_{ult}(T)$ is the ultimate tensile stress, σ_{appl} is the applied tensile fatigue stress, $E_c(T)$ is Young's modulus of the composite, n is the percentage drop in stiffness, F is a term related to the volume fraction of fiber and off-axis angle, T is the temperature effect, σ_{max} is the maximum fatigue stress, and R is the stress ratio. The D–N curve parameters of basalt fiber-reinforced polymer composites are listed in Table 7.7 [56].

The weakest point effect can be obtained under fatigue load in this case study at the inner wall of the pipe, as the maximum temperature at the inner pipe wall equals $T_I = 333.15$ K and the minimum temperature at the outer pipe wall equals $T_O = 293.15$ K. Figure 7.21 shows the D–N curve of basalt fiber-reinforced polymer. As shown in the figure, the failure mode of the pipeline satisfies the cumulative fatigue damage index rule described in formula (7.9). As shown in Figure 7.22, the fatigue damage increases significantly when the melting point of the matrix is approached.

Table 7.7: D–N curve parameters of basalt fiber-reinforced polymer composites.

Parameter	Basalt fiber-reinforced polymer
N_f	6.12×10^4
E_m (GPa)	4.06
V_m	0.43
E_f (GPa)	97
V_f	0.53
E_c (GPa)	84
T_m (K)	446
f^*	0.52
S_f (MPa)	137
S_{ult} (MPa)	1,409
σ_{min} (MPa)	14.2
σ_{max} (MPa)	139
R	0.10
n	1.66

Note: These parameters can be found in reference [56].

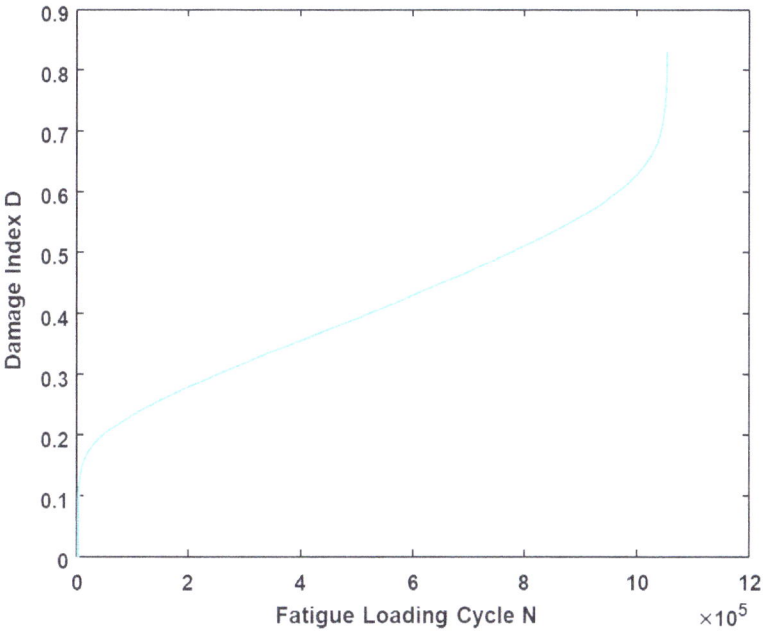

Figure 7.21: D–N curve of basalt fiber-reinforced polymer.

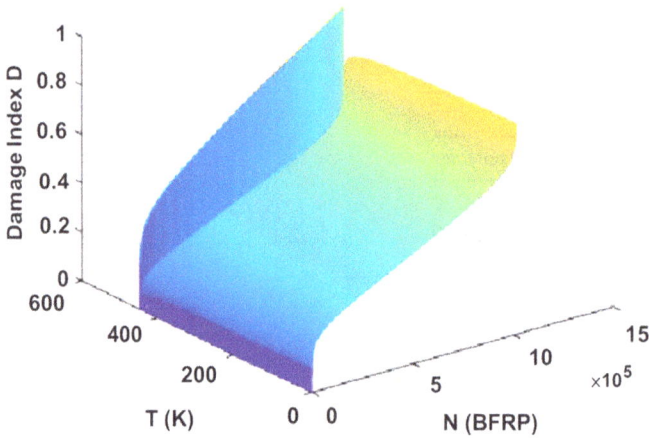

Figure 7.22: *D–T–N* relationship of basalt fiber-reinforced polymer.

7.1.3.4 Deep neural networks

7.1.3.4.1 Deep neural network configuration for LTCTMF behavior in basalt fiber-reinforced polymer composite pipeline

To predict the LTCTMF behavior in basalt fiber-reinforced polymer composite pipelines, a new deep neural network is established from a multilayer neural network to get a promising deep neural network architecture capable of predicting the LTCTMF behavior in different composite structures.

Figure 7.23 illustrates the proposed deep neural network of fully connected layers. The consecutive hidden layers of the deep neural network are positioned between the input layer (features) and the output layer (object predictions) to enable advanced feature extraction through the training process. Using weights W and biases b, the deep neural network is trained to identify more useful nonlinear information.

The features from electrical capacitance sensor (finite element model) data are extracted using the average pooling operations in one-dimensional convolution. There are 6 hidden layers included in the fully connected neural network, the first 3 layers have 502 neurons, and the last 3 layers have 251 neurons. ReLUs are used as activation functions for both the convolutional and fully connected layers. In the proposed deep neural network, this process is defined in Equation (7.2).

The definitions of the equation's symbols can be found in the table of annotations in the Appendix at the end of the case study.

A suitable value for the weights and the bias of the hidden layers can be determined as follows:

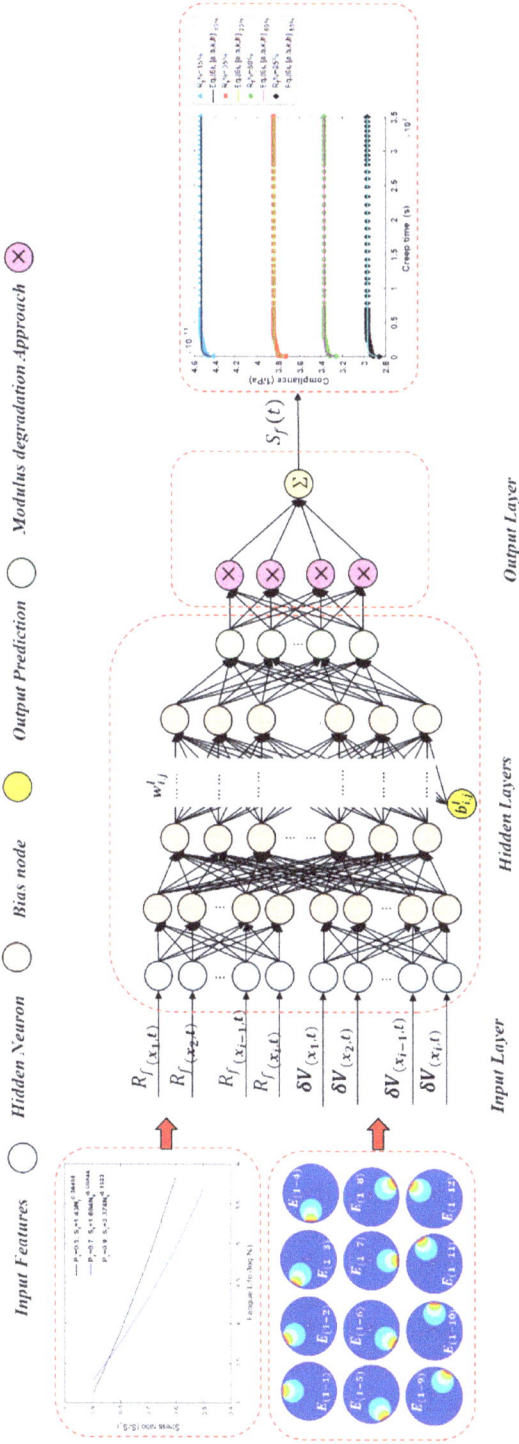

Figure 7.23: The architecture of the proposed deep neural network to predict LTCTMF compliance ($S_f(t)$) with input, sequential hidden, and output layers.

$$w = (\text{Number of input variables} + 1) \times \text{Number of neurons} \qquad (7.10)$$

And each input of the hidden layers will have a bias term equal to the "number of layer input variables + 1." (*Number of layer input variables* + 1)

The architecture of the CNN layers used in this case study is shown in Figure 7.9, including a subsampling layer (S1), and a convolution layer (C1), which comprises six feature maps. The details of the subsampling and convolution layers are provided in Table 7.8.

Table 7.8: Detailed settings of the convolutional layers.

Layer	Kernel size	Number of parameters	Number of connections
C1	12×12	234	540,136
S1	8×12	8	68,632
C2	12×12	//	//
S2	8×8	//	//

From Table 7.8, the size of the input matrix connection for each feature map neuron is 12×12, while the matrix size of the feature map is 156×156. For the filters used in the proposed deep neural network, a total of 4 filters are utilized, each having 12×12 unit parameters for weights and biases, resulting in a total of $(12 \times 12 + 1) \times 8 = 1{,}160$ tuning parameters. The kernel is used between the input and the first convolutional layers C1; therefore, the total number of connections is $1{,}160 \times (156 \times 156) = 28{,}229{,}760$. A subsampling layer S1 has $(2 + 2) \times 2 = 8$ tuning parameters and $(8 \times 8 + 1) \times 4 \times (112 \times 112) = 3{,}261{,}440$ connections. The next subsampling layer S2 has a similar architecture (see Table 7.8).

7.1.3.4.2 Deep neural network training and testing

The datasets used for training and testing the proposed deep neural network are obtained from finite element model results by ANSYS, which are presented in Section 7.1.3.8 for predicting the LTCTMF behavior in basalt fiber-reinforced polymer composite pipelines.

The algorithm used for training and testing the deep neural network, which incorporates case studies with adaptive features and variable parameters, is illustrated in Figure 7.24.

First, the hidden layers are controlled by setting the parameters based on the finite element model datasets. The process takes place in two stages, as shown in Figure 7.24, training (actual) and testing (expected), with labeled data on the right and left, respectively. The number of hidden layers can change depending on the results of the training phase. The actual data appears after processing the results. It is expected that the testing process will be performed multiple times (iterations) until satisfactory results are achieved. The steps of deep neural network settings for training and testing models are shown in Figure 7.25.

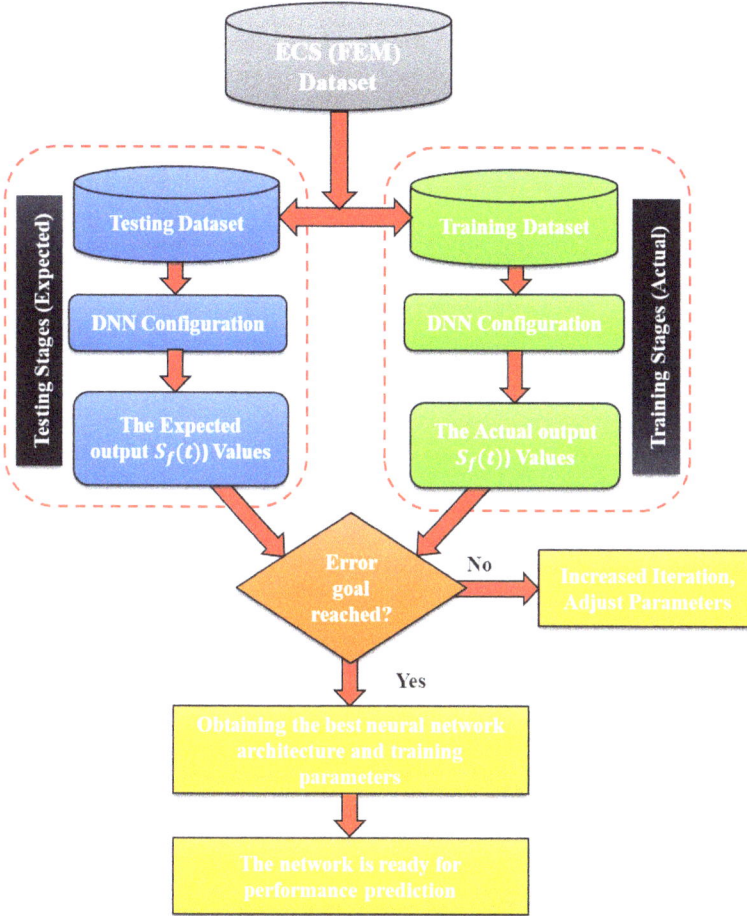

Figure 7.24: The flowchart of LTCTMF compliance $(S_f(t))$ prediction processes.

There are 78,114 specimens in the finite element model data of LTCTMF compliance $(S_f(t))$ over the time of creep, which are fed into the deep neural network model for training. The total number of training specimens is divided into two parts as follows: 80% are randomly selected as the training set and the remaining 20% serve as the test set. Then use the pretrained network for transfer learning, rotate all the data for 500 epochs, and use the mean squared error as the loss function (see Figure 7.17). The total batch time for training and validation is 63 s and 5.22 s, respectively, and the training key parameters of training rate and attenuation factor are $1E-4$ and $1E-6$, respectively.

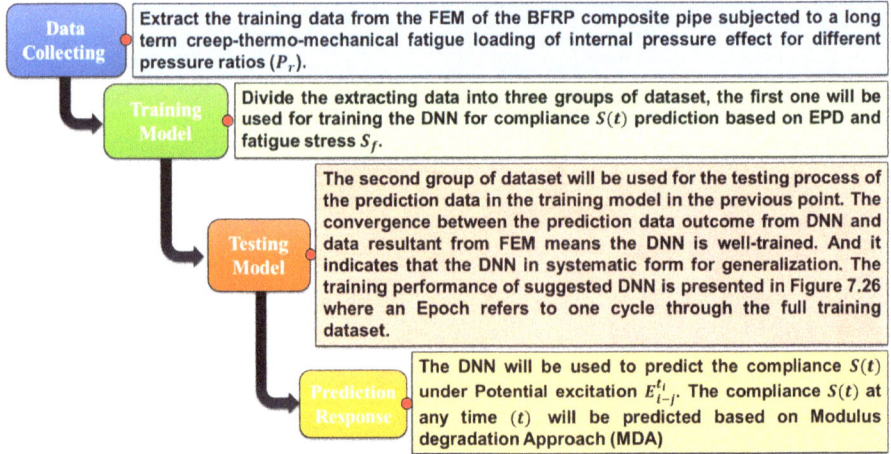

| Data Collecting | Extract the training data from the FEM of the BFRP composite pipe subjected to a long term creep-thermo-mechanical fatigue loading of internal pressure effect for different pressure ratios (P_r). |

| Training Model | Divide the extracting data into three groups of dataset, the first one will be used for training the DNN for compliance $S(t)$ prediction based on EPD and fatigue stress S_f. |

| Testing Model | The second group of dataset will be used for the testing process of the prediction data in the training model in the previous point. The convergence between the prediction data outcome from DNN and data resultant from FEM means the DNN is well-trained. And it indicates that the DNN in systematic form for generalization. The training performance of suggested DNN is presented in Figure 7.26 where an Epoch refers to one cycle through the full training dataset. |

| Prediction Response | The DNN will be used to predict the compliance $S(t)$ under Potential excitation $E_{t-f}^{t_i}$. The compliance $S(t)$ at any time (t) will be predicted based on Modulus degradation Approach (MDA) |

Figure 7.25: Illustration of deep neural network training steps to predict LTCTMF compliance ($S_f(t)$).

Figure 7.26 shows the train and test mean squared error loss using supervised mode. It can be seen that as the training epochs increase, the mean squared error decreases to a small value, and the loss related to $S_f(t)$ shows a small value and slowly decreases after 80 epochs. Moreover, the test (validation) mean squared error has loss values

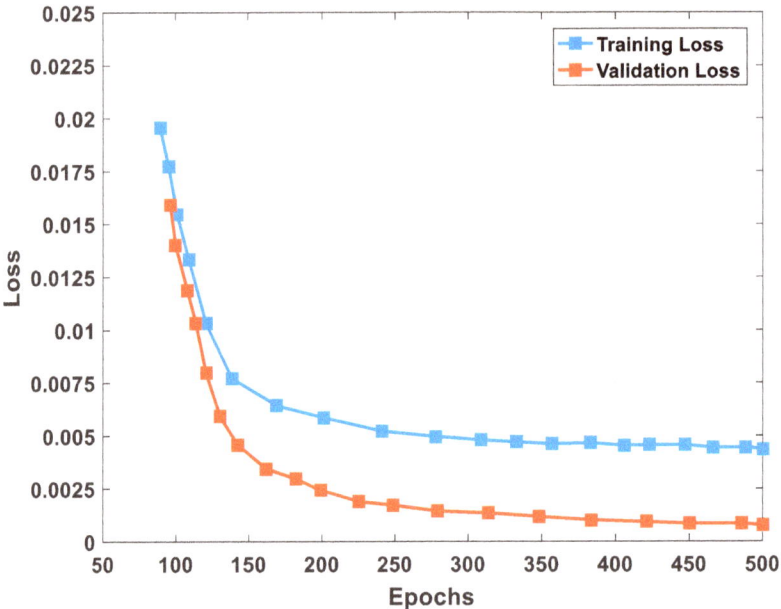

Figure 7.26: The proposed deep neural network training and testing of mean squared error loss.

larger than the training mean squared error values, while both almost reach a steady state at the same epoch. Overall, better performance in training and testing can be observed for predicting LTCTMF behavior in the basalt fiber-reinforced polymer composite pipeline.

7.1.3.4.3 The training and testing algorithms

Algorithm 7.2 lists deep neural network training and testing steps, and the neural network is developed and implemented using MATLAB software.

Algorithm 7.2: Training and testing of the proposed deep neural network.

1: **input**: d: creep compliance ($S(t)$) dataset, I: permittivity ε, Creep time, EPC, W: Network parameter matrix weight w_{ij} and bias b_j

2: **output**: score of deep neural network trained model on the dataset to predict $S(t)$ for various ε, EPD, Creep time

3: **let** f be the feature set 3d matrix

4: **for** i in the dataset **do**

5: **let** f_i be the feature set matrix of sample l

6: **for** j in i **do**

7:   $V_i \leftarrow vectorize_{(j,w)}$

8:   **append** V_i to f_i

9:   **append** f_i to f

10:   $f_{train}, f_{test}, l_{train}, l_{test} \leftarrow$ the split feature set and prediction into train subset and test subset

11:   $M \leftarrow$ deep neural network (f_{train}, l_{train})

12:   score $\leftarrow$ evaluation (l, , l_{test}, M)

13:   **return** score

14: **end for**

15: **end for**

7.1.3.5 Electrical capacitance sensor results for LTCTMF behavior in basalt fiber-reinforced polymer composite pipelines

Twelve electrodes are installed on the outer surface of the pipe specimen. For various levels of LTCTMF ($R_f\%$) equal to (25%, 50%, and 75%), the electric potential difference between the electrical capacitance sensor electrode pairs is measured. Based on the measurement data, the LTCTMF compliance ($S_f(t)$) over the duration of creep is analyzed using ANSYS software.

Using the scripting capabilities in ANSYS, we apply the basic functionalities of the modulus degradation approach to describe the LTCTMF compliance $S_f(t)$ data in MATLAB code. This will assist us in developing scripts in ANSYS Discovery for the simulation of LTCTMF behavior. The finite element model data plotting of $S_f(t)$ over the time of creep is presented in Figure 7.27.

From Figure 7.27, it can be observed that the results of LTCTMF compliance $(S_f(t))$ against creep time show that the tendency curve for $R_f\% = 25\%$ has the highest $S_f(t)$, and the curve for $R_f\% = 75\%$ was given the lowest $S_f(t)$, while the curve for $R_f\% = 50\%$ was laid between them. The variance in $S_f(t)$ between different $R_f\%$ was due to fluctuations in the composite material modulus over the creep time. In this context, the reductions of $S_f(t)$ over creep time equal 35,000,000 s at $R_f\% = 25\%$ and 75% were estimated at 20% and 68%, respectively.

A built-in MATLAB curve-fitting routine is used to fit the finite element model data of $S_f(t)$. The routine employs a least-squares algorithm to optimize a set of design parameters that minimize the sum of the squares of the errors. By using the exponential formula (7.11), it was found to be suitable, yielding a high CF near unity, equal to 0.9863, 0.9452, 0.9671 for $R_f\% = 25\%$, 50%, and 75%, respectively. The constants of that formula, known as the LTCTMF compliance constants, are listed in Table 7.9 for each $R_f\%$:

$$S_f(t) = ae^{bt} + ke^{ht} \tag{7.11}$$

Table 7.9: LTCTMF compliance constants.

LTCTMF ($R_f\%$)	CF	$a \times 10^{-11}$	$b \times 10^{-20}$	$k \times 10^{-13}$	$h \times 10^{-6}$
25%	0.9863	4.33	8.337	−10.45	−3.333
50%	0.9452	3.577	8.016	−12.249	−2.92
75%	0.9671	3.16	8.51	−13.231	−2.886

Figures 7.28 and 7.29 show the effects of changes in the LTCTMF level ($R_f\%$) change and the duration of creep, respectively, on the responses of electrical capacitance sensors (3D ANSYS model) across 21 capacitance measurements (C_{ij}).

As shown in Figures 7.28 and 7.29, the electrical capacitance sensor responses are affected by $R_f\%$ and creep time changes. It found a significant degradation in electrical capacitance sensor responses of capacitance with an increase of $R_f\%$ levels and creep time elapsed:

$$\text{ECS sensitivity } \% = \frac{C_0 - C_t}{C_0} \times 100 \tag{7.12}$$

where C_0 and C_t are the capacitance measurements for LTCTMF at creep time equal to zero and during LTCTMF monitoring time, respectively.

As shown in Figure 7.30, the sensitivity of the electrical capacitance sensors increases with creep time increases at the first 20% of creep time until the stability of sensitivity occurs at almost 73% with the remainder 80% of the creep time.

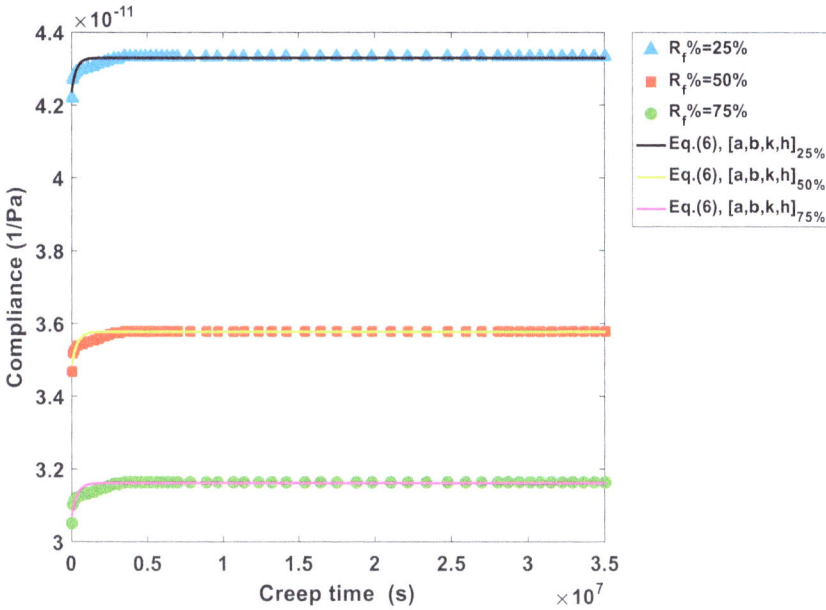

Figure 7.27: The LTCTMF compliance $(S_f(t))$ at $R_f\% = 25\%, 50\%, 75\%$.

7.1.3.6 Validity of the proposed technique

7.1.3.6.1 Theoretical validation

The modulus degradation approach is used to compute the modulus reduction in the resin of composite materials during creep. The instantaneous modulus of the composite material's resin can be determined as follows:

$$E(t) = \frac{\sigma(t)}{\varepsilon(t)} \tag{7.13}$$

Therefore, the strain variations over time, which correspond to the reduction in resin modulus, can be simulated. This is a promising approach for evaluating the relative strain between the fiber and resin during creep by partitioning the applied force between them. By assuming that the strains in the matrix and fiber are identical under the iso-strain condition at the microscale, we can simulate the representative volume element as follows:

$$\varepsilon = \varepsilon_{\text{Fiber}} = \varepsilon_{\text{Matrix}} \tag{7.14}$$

By applying the equilibrium equation under the iso-strain condition, the following can be obtained:

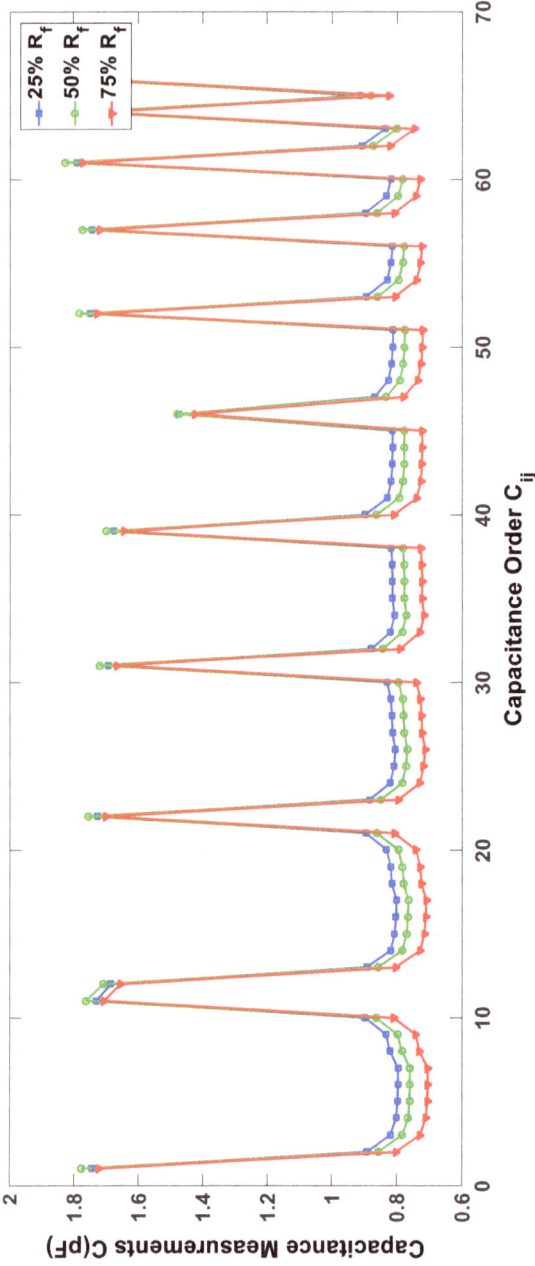

Figure 7.28: The LTCTMF (R_f%) effect on electrical capacitance sensor capacitance response.

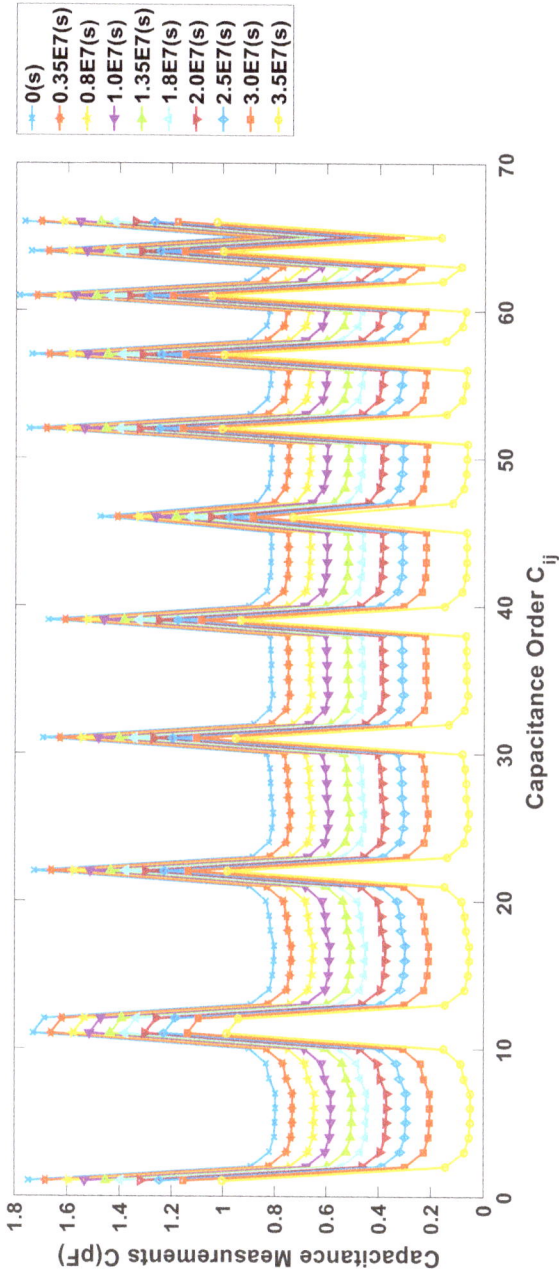

Figure 7.29: The creep time effect on electrical capacitance sensor capacitance response.

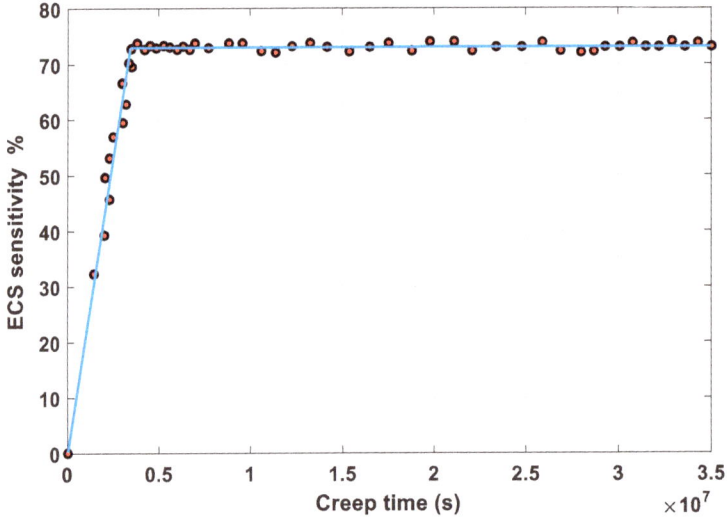

Figure 7.30: Sensitivity of electrical capacitance sensors versus creep time.

$$\sigma = \sigma_{\text{Fiber}} V_{\text{Fiber}} = \sigma_{\text{Matrix}} V_{\text{Matrix}} \tag{7.15}$$

Thus we have:

$$\sigma_{\text{Matrix}} = \frac{\sigma}{\frac{E_{\text{Fiber}}}{E_{\text{Matrix}}} V_{\text{Fiber}} + V_{\text{Matrix}}} \tag{7.16}$$

where V is the volume fraction; therefore, by taking each stress along the fiber direction, the stress in the resin can be calculated using Equation (7.16).

During multistage loading, the creep compliance function over the time of creep can be expressed as follows [422]:

$$\varepsilon(t) = \sigma_1 D(t - \tau_1) + \sigma_2 D(t - \tau_2) + \sigma_3 D(t - \tau_3) + \cdots \tag{7.17}$$

where $D(t - \tau_i)$ is the creep compliance function. By using Boltzmann's law [67] for simulating the continuous variation of stress in the integral form as follows:

$$\varepsilon(t) = D_0 \sigma + \int_0^4 \Delta D(t - \tau) \frac{d\sigma}{d\tau} dt \tag{7.18}$$

The material compliance in linear viscoelastic behavior can be considered a function of time based on stress status; however, in nonlinear viscoelastic behavior, it is considered a function of both time and stress variables. As demonstrated by Boltzmann's law, material compliance is divided into initial (D_0) and transient ($D(t)$) terms. Transient compliance is usually obtained from the curve-fitting law of finite element model data using software of curve-fitting routine, or through the Prony series representation of the creep compliance function [433].

In this case study, the transient compliance term in Boltzmann's law (Equation (7.16)) for LTCTMF is derived using the curve-fitting law in Equation (7.11), and various finite element models through changes of $R_f\%$ values are 25%, 50%, and 75%. The constants in Equation (7.11) are obtained from Table 7.9. Figure 7.31 shows a comparison between the electrical capacitance sensor (finite element model) data and theoretical data for LTCTMF compliance $(S_f(t))$ at $R_f\% = 25\%$, 50%, 75%. As shown in the figure, the modulus degradation approach data gave excellent agreement with the finite element model data of $S_f(t)$.

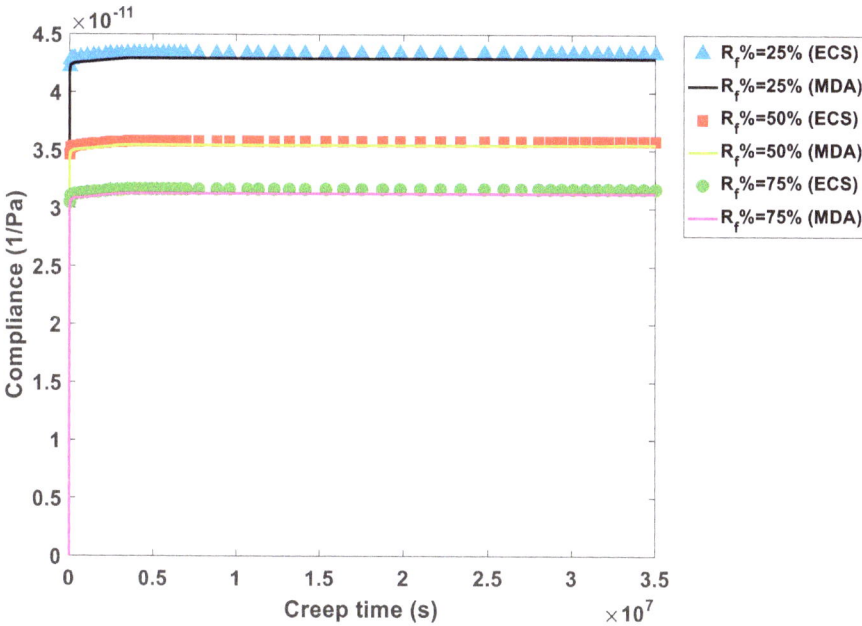

Figure 7.31: A comparison between electrical capacitance sensor data (finite element model) and modulus degradation approach data (theoretical) for LTCTMF compliance $(S_f(t))$ at $R_f\% = 25\%, 50\%, 75\%$.

Figure 7.32 shows the evaluation of the correlation between the results obtained from electrical capacitance sensors (finite element model) and theoretical results using the modulus degradation approach for $S_f(t)$ at $R_f\% = 25\%$, 50%, 75%. As shown in the figure, the correlation between the finite element model and the theoretical data has a high CF equal to 0.9999, which is very close to unity. The maximum error between the finite element model and the theoretical data, referred to as the error band from the diagonal line in the figure, is less than $0.45 \times 10^{-12}\,\mathrm{Pa}^{-1}$ for $S_f(t)$. The theoretical data gave excellent estimations for the finite element model data for $S_f(t)$ in the composite pipeline.

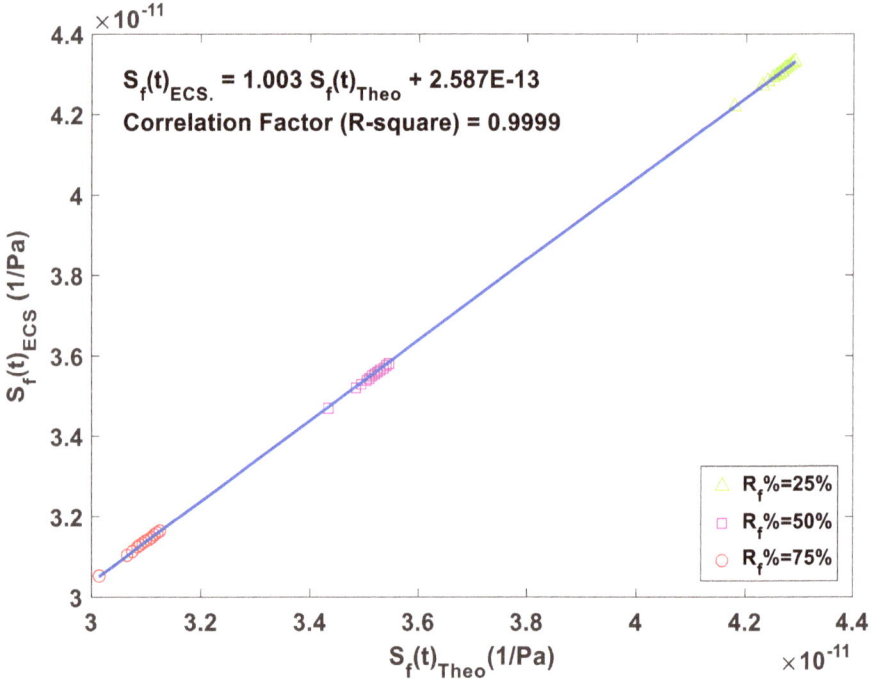

Figure 7.32: The correlation between the results of electrical capacitance sensors (finite element model) and modulus degradation approach (theoretical).

7.1.3.6.2 Experimental validation

In this case study, the experimental dataset adapted from Rafiee and Ghorbanhosseini [424, 425] is used to validate the proposed technique. The long-term creep (LTC) tests of glass fiber-reinforced polymer composite pipes were performed under single-stage loading (SSL). The specimens have 12 m length and 500 mm diameter, and the fiber ply stacking sequence is $[90_2/\pm 60.19_4/90]$, and each hoop layer was measured as 1.5 mm, 0.44 mm, and 0.38 mm, respectively. The manufacturing process used is reciprocal filament winding, commercially referred to as discontinuous filament winding technology. The fiber and polyester resin volume fractions (V_f) were measured at 56% and 44%, respectively. The test specimens were prepared in complete ring form and were cut from the produced full-length pipe with a length of 300 mm.

The LTC test of the pipe was performed based on ISO7684 [426] and the appropriate device was manufactured in situ. Rafiee and Ghorbanhosseini [424, 425] conducted the creep tests under an SSL equal to 981 N on ring specimens. The time of creep for the load applied to the specimen was 35,000,000 s. The test setup, initiation, and execution of the LTC test on the ring specimen are presented in Figure 7.33.

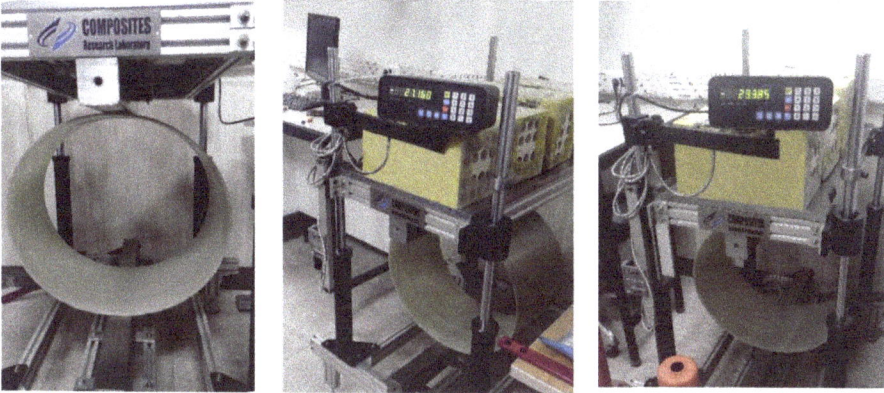

Figure 7.33: The creep test setup on the specimen using the LTC test device [424, 425].

To perform a convergence investigation for the proposed technique, a finite element model was established using the same material properties and geometric specifications as the pipe sample used in the experimental work conducted by Rafiee and Ghorbanhosseini [424, 425]. The finite element model was utilized to detect the LTC under SSL by installing 12 electrodes of electrical capacitance sensors on the outer periphery of the pipe and exciting it by $V_0 = 15$ V. The electric potential difference between electrode pairs was measured, corresponding to LTC compliance ($D_{\mathrm{LTC}}(t)$) over the time of creep, and was compared with the experimental results of Rafiee and Ghorbanhosseini [424, 425] as well as theoretical results obtained via the modulus degradation approach, as shown in Figure 7.34.

Figure 7.35 shows the evaluation of the correlation between the results from electrical capacitance sensors (finite element model) and experimental data, as adapted by Rafiee and Ghorbanhosseini [424, 425] for $D_{\mathrm{LTC}}(t)$. As shown in the figure, the correlation between the finite element model and experimental data has a high CF equal to 0.9796, which is very close to unity. The error band from the diagonal line in the figure is less than 2.7×10^{-12} Pa^{-1} for $D_{\mathrm{LTC}}(t)$. The finite element model gave excellent estimations of the experimental data for $D_{\mathrm{LTC}}(t)$ in composite pipes.

7.1.3.7 The deep neural network predicted data of $S_f(t)$

The deep neural network configuration in this research is shown in Figure 7.23. The deep neural network is trained to predict LTCTMF compliance ($S_f(t)$) by measuring values of $R_f\%$, creep time, and electric potential difference in the input layer. The electric potential difference measurements were extracted between electrical capacitance sensor electrodes for LTCTMF levels at $R_f\% = 25\%$, 50%, and 75% numerically under the pipeline's initial conditions are internal pressure of $P_0 = 1{,}500E4$ Pa, and a temperature difference between the inside and outside of the pipe wall of $\Delta\theta = 40°$.

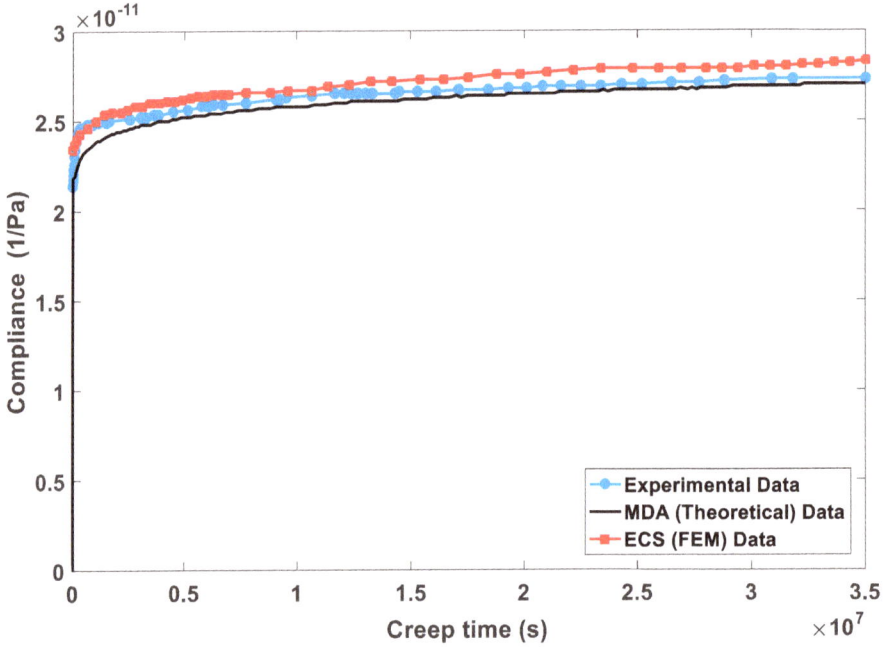

Figure 7.34: A comparison between electrical capacitance sensor (finite element model) data, the modulus degradation approach (theoretical) data, and experimental data by Rafiee and Ghorbanhosseini [424, 425] for $D_{LTC}(t)$.

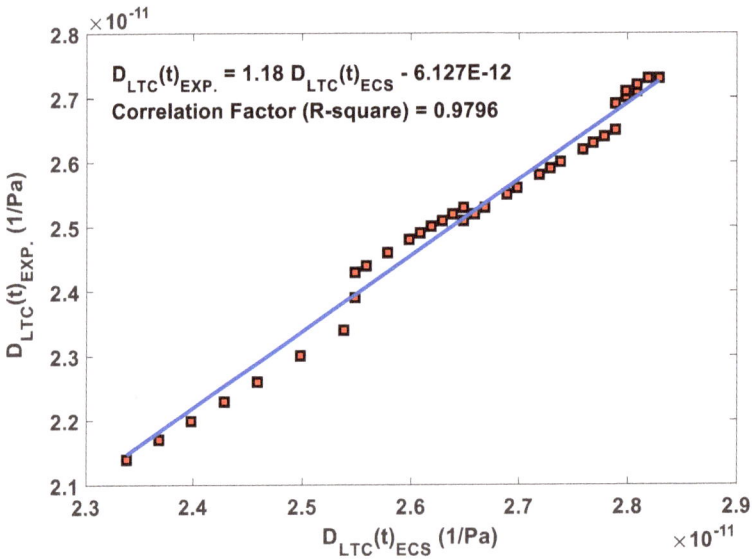

Figure 7.35: The correlation between experimental and electrical capacitance sensor (finite element model) results.

Figure 7.36 shows a comparison between the electrical capacitance sensor (finite element model) data and deep neural network predicted data for LTCTMF compliance $(S_f(t))$ at $R_f\% = 25\%,\ 50\%,\ 75\%$.

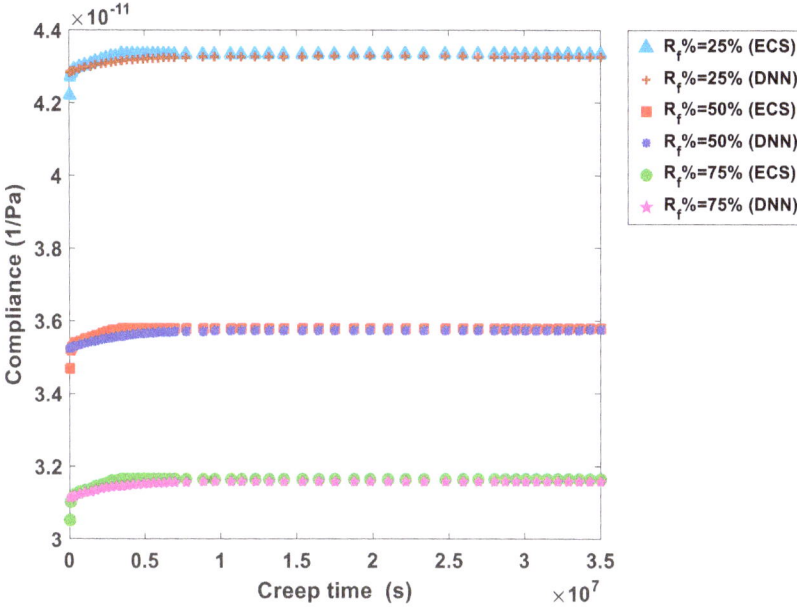

Figure 7.36: A comparison between the electrical capacitance sensor (finite element model) data and deep neural network predicted data for LTCTMF compliance $(S_f(t))$ at $R_f\% = 25\%, 50\%, 75\%$.

As shown in the Figure 7.36, deep neural network data gave excellent agreement with finite element model data regarding LTCTMF compliance with high accuracy in predictions and low mean squared error values between the finite element and deep neural network predicted results (see Equation (7.19)) as presented in Table 7.10:

$$\text{MSE} = \sum \left((E_{i-j})_{\text{DNN}} - E_{i-j} \right)^2 / n \tag{7.19}$$

where $(E_{i-j})_{\text{DNN}}$ is the predicted electric potential difference, E_{i-j} is the electric potential difference measured between the electrical capacitance sensor electrodes, and n is the number of finite element model data.

Figure 7.37 shows the evaluation of the correlation between the electrical capacitance sensor (finite element model) results and deep neural network predicted results for $S_f(t)$ at $R_f\% = 25\%,\ 50\%,\ 75\%$. As shown in the figure, the correlation between the finite element model and the predicted data exhibits a high CF of 0.9899, which is very close to unity. The error band from the diagonal line in the figure is less than $2 \times 10^{-12} \frac{1}{Pa}$ for $S_f(t)$. The deep neural network provided excellent estimations of the finite element model data for LTCTMF compliance in the composite pipeline.

Table 7.10: Mean squared error values.

Data	LTCTMF (R_f%)	Mean squared error
Deep neural network-electrical capacitance sensors (DNN-ECS)	25%	1.4867e-26
	50%	1.3807e-22
	75%	5.8019e-23

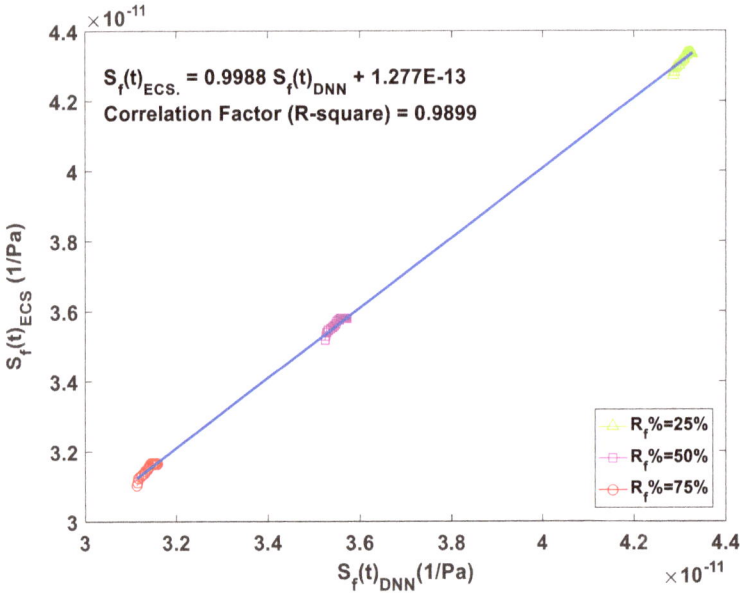

Figure 7.37: The correlation between electrical capacitance sensors (finite element model) and deep learning results.

7.1.3.8 Utilizing the trained deep neural network for predicting nonfinite element model data

In this stage, the trained deep neural network configuration described in Section 7.1.2.4 is used to predict LTCTMF behavior for other new LTCTMF levels (R_f%) not extracted by the electrical capacitance sensor technique (nonfinite element model data) equal 15%, 35%, 60%, and 85%. The LTCTMF compliance ($S_f(t)$) is predicted under the same initial conditions, where the internal pressure of the pipeline is $P_0 = 1,500E4$ Pa, and the temperature difference between the inside and outside pipe wall is $\Delta\theta = 40°$.

Figure 7.38 shows the predicted nonfinite element model results of the LTCTMF compliance $(S_f(t))$ in the basalt fiber-reinforced polymer composite pipeline.

From Figure 7.38, it can be observed that the results of LTCTMF compliance $(S_f(t))$ against creep time show that the tendency curve for $R_f\% = 15\%$ has the highest $S_f(t)$, and the curve for $R_f\% = 75\%$ was given the lowest $S_f(t)$, while the other $R_f\%$ curves were laid between them. The variance in $S_f(t)$ between different $R_f\%$ was due to fluctuations in the composite material modulus over the creep time. In this context, the reductions of $S_f(t)$ over a creep time equal 35,000,000 s at $R_f\% = 15\%$ and 85% were estimated at 216% and 78%, respectively.

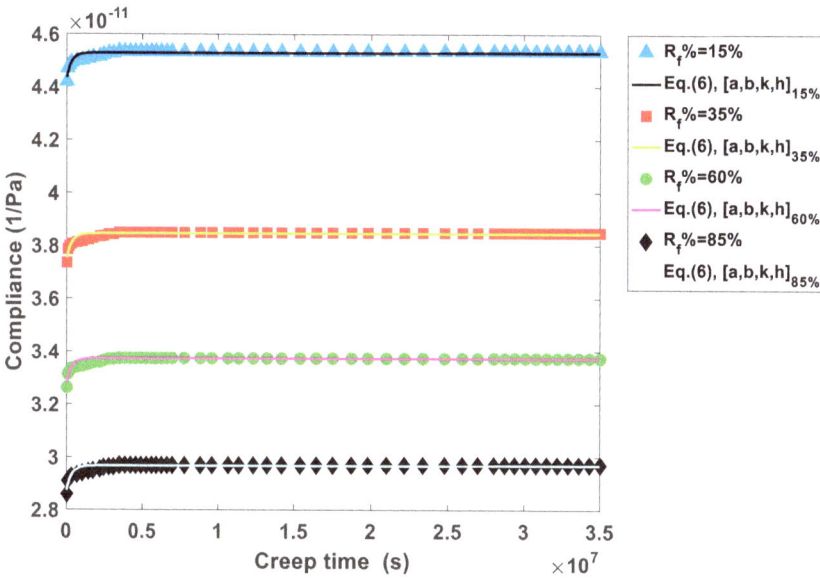

Figure 7.38: The nonfinite element model predicted data for LTCTMF compliance $(S_f(t))$ at $R_f\% = 15\%, 35\%, 60\%, 85\%$.

Employing the exponential formula (7.10) to fit the nonfinite element model results of $S_f(t)$ has proven to be suitable, as it provides an acceptable value for the CF, as presented in Table 7.9. The LTCTMF compliance constants are listed in Table 7.11 for each $R_f\%$, indicating that the proposed deep neural network is both suitable and useful in predicting nonfinite element model data.

Table 7.11: Creep compliance constants a, b, k, and h for tensile creep (%σ_c).

LTCTMF (R_f%)	CF	$a \times 10^{-11}$	$b \times 10^{-20}$	$k \times 10^{-13}$	$h \times 10^{-6}$
**15%	0.9247	4.529	8.019	−9.417	−3.262
*25%	0.9863	4.33	8.337	−10.45	−3.333
**35%	0.9577	3.844	8.298	−11.258	−2.938
*50%	0.9452	3.577	8.016	−12.249	−2.92
**60%	0.9381	3.371	8.606	−12.676	−2.973
*75%	0.9671	3.16	8.51	−13.231	−2.886
**85%	0.922	2.967	8.221	−13.861	−2.945
Avg			8.286714		−3.036714
SD			0.224993		0.1812417

*Finite element model data. **Deep neural network data.

7.1.3.9 Summary

This case study adopted a new framework for monitoring LTCTMF behavior in basalt fiber-reinforced polymer laminated composite pipelines by measuring the electric potential difference between electrical capacitance sensor electrodes due to changes in the dielectric permittivity of the pipeline material. The composite pipe was subjected to long-term fatigue loading caused by internal pressure and the thermal effects of temperature differences between the inside and outside pipe wall temperatures. The internal pressure was $P_0 = 1{,}500E4$ Pa, and the liquid inside the pipe had a temperature of 60 °C, while the outside temperature was 20 °C. The electric potential difference measurements were extracted between electrical capacitance sensor electrodes for various LTCTMF levels R_f% = 25%, 50%, and 75% numerically analyzed to train a novel deep neural network architecture designed to predict LTCTMF behavior for other variations of R_f% = 15%, 35%, 60%, and 85%. The results of the proposed method were verified theoretically using a modulus degradation approach to simulate LTCTMF in the pipeline material and experimentally using the experimental results available in the literature. The results obtained are as follows:

1) The integration of electrical capacitance sensors and deep learning approaches can serve as a promising method for monitoring LTCTMF behavior in composite structures.

2) The results of the proposed approach showed excellent agreement with theoretical, experimental, and predicted outcomes, as illustrated in Figures 7.32, 7.35, and 6.37, respectively.

3) The results of LTCTMF compliance ($S_f(t)$) against creep time for all levels of LTCTMF (R_f%) show that the tendency curve for R_f% = 15% has the highest $S_f(t)$, and the curve for R_f% = 85% was the lowest $S_f(t)$, while the other R_f% curves were laid between these extremes in descending order from R_f% = 25% to R_f% = 75%. The variance in $S_f(t)$ between different R_f% was attributed to fluctuations in the composite material modulus over the CT. In this regard, the reductions

of $S_f(t)$ over a creep time equal 35,000,000 s at $R_f\% = 15\%$ and 85% were estimated at 16% and 78%, respectively.

4) The exponential formula $S_f(t) = ae^{bt} + ke^{ht}$ demonstrates its applicability to this study. It was found that for different $R_f\%$ and $S_f(t)$, the deviation of the constant (b and h) is negligible and can be considered constant, as the corresponding standard deviation (SD) was found to have acceptable values, as shown in Table 7.11.

5) Constants (a and k) are defined as the creep constants because they were found to be associated with the $S_f(t)$ values according to the $R_f\%$ with a high CF, and they decrease in descending order from $R_f\% = 15\%$ to $R_f\% = 85\%$, that is, as the $S_f(t)$ increases.

6) Finally, it can be concluded that the proposed method provides a better understanding of the behavior of LTCTMF under different $R_f\%$.

7.1.4 Case study (3): structural health monitoring of composite pipelines utilizing fiber optic sensors and machine learning approach

In this case study, a structural health monitoring system is proposed to provide automatic early warning for detecting damage and its location in composite pipelines at an early stage. The study considers a basalt fiber-reinforced polymer pipeline with an embedded fiber Bragg grating (FBG) sensory system and first discusses the shortcomings and challenges of incorporating FBG sensors for accurate detection of damage information in pipelines. The novelty and main focus of this study, however, is a proposed approach that relies on designing an integrated sensing-diagnostic structural health monitoring system capable of detecting damage in composite pipelines at an early stage via implementation of an artificial intelligence-based algorithm that combines deep learning and other efficient machine learning methods using an enhanced CNN, without retraining the model. The proposed architecture replaces the softmax layer with a k-nearest neighbor algorithm for inference. Finite element models are developed and calibrated using the results of pipe measurements under damage tests. These models are then used to assess the patterns of strain distributions in the pipeline under internal pressure loading and pressure changes due to bursts, as well as to determine the relationship of strains at different locations axially and circumferentially. A prediction algorithm for pipe damage mechanisms based on distributed strain patterns is also developed. The enhanced CNN is designed and trained to identify the condition of pipe deterioration, so the initiation of damage can be detected. The strain results from the current method show excellent agreement with available experimental results in the literature. The average error between the enhanced CNN data and FBG sensor data is 0.093%, thus confirming the reliability and accuracy of the proposed method. The proposed enhanced CNN achieves high performance with 93.33% accuracy ($P\%$), 91.18% regression rate ($R\%$), and a 90.54% F1-score ($F\%$).

7.1.4.1 Methodology

The methodology used in this research aims to detect damage and assess the safety of composite pipeline structures through a highly effective and reliable approach. It then integrates this approach into a practical structural health monitoring system that is suitable for both old and new piping system operations.

More specifically, this case study aims to:

Figure 7.39 illustrates the stages of the proposed approach. The aim is to assess the feasibility of developing and operating a structural health monitoring system for the early detection of damage in composite pipelines. The flowchart outlining the proposed method is shown in Figure 7.39.

More specifically, this case study aims to:

(1) Establish an understanding of the characteristics of FBG sensors in a composite structure made of basalt fiber-reinforced polymer composite pipe.
(2) Assess the effects of damage occurring in the basalt fiber-reinforced polymer composite pipeline on the displacement response from the dynamic signal for each FBG, including its impact on dispersion, attenuation, and scattering.
(3) Develop a sensory health monitoring platform for the early detection and classification of damage in composite pipelines.
(4) Propose a novel artificial intelligence-based structural health monitoring system that integrates with FBG sensors to provide damage classification in composite pipelines.
(5) Suggest a hybrid approach to improve the accuracy of automatically identifying damage in pipelines by combining deep learning and machine learning in a new algorithm. This approach utilizes a hybrid CNN + k-nearest neighbor algorithm and known in this case study as ECNN.
(6) Verify the effectiveness and accuracy of the proposed artificial intelligence algorithm based on four indices: TPR, TNR, FPR, and FNR. Additionally, the accuracy rate (P), regression rate (R), and F1-score (F) can be determined for the proposed artificial intelligence algorithm.

7.1.4.2 The geometric model

The finite element model of a composite pipeline is established using ANSYS software. Both free meshing and tetrahedral meshing are applied. The minimum element size and element type for the composite pipeline are selected as 0.005 m for element size and SHELL 181 for element type, respectively. The composite pipeline dimensions are 1 m long, and the distribution of which is $[-45, 45, -45, 45, -45]$. The density of the model is 2.8 g/cm^3. The internal radius (r_i) is 0.04 m. The external radius (r_o) is 0.043 m, and the length (L) is 1.0 m. Both ends of the pipe are fixed as boundaries. Table 7.12 provides a detailed list of the mechanical parameters of the basalt fiber-reinforced polymer composite pipeline properties, where E_x, E_y, E_z are the elastic moduli in the "x", "y," and "z" directions, G_{xy}, G_{xz}, G_{yz} are the shear moduli in the "xy," "xz," and "yz" planes, respectively; and v_{xy}, v_{xz}, v_{yz} are the Poisson's ratios in the "xy," "xz," and "yz" planes, respectively.

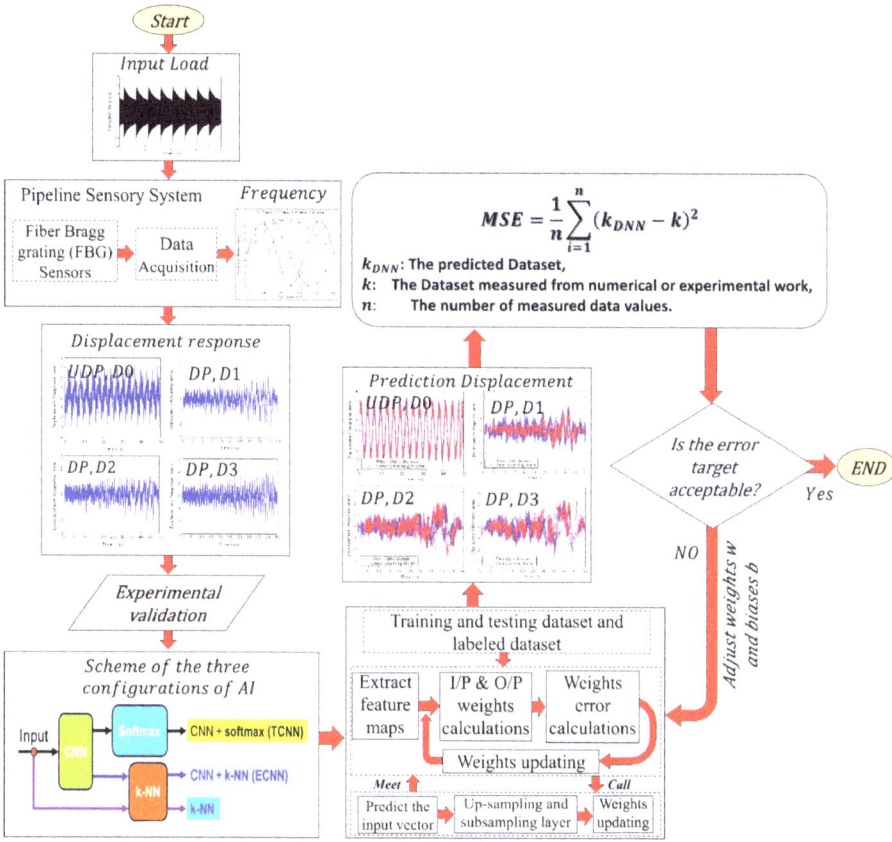

Figure 7.39: Methodology flowchart.

Table 7.12: Structural properties of basalt fiber-reinforced polymer composites.

E_x (Pa × 10^9)	93.5
E_y (Pa × 10^9)	20
E_z (Pa × 10^9)	20
G_{xy} (Pa × 10^9)	8.5
G_{yz} (Pa × 10^9)	2.35
G_{xz} (Pa × 10^9)	2.35
v_{xy}	0.28
v_{yz}	0.30
v_{xz}	0.28

7.1.4.2.1 Damaged pipeline system modeling

The damage is modeled by a reduction in the stiffness of the pipeline. An enlarged view of the damaged areas (marked in purple) on the pipeline is shown in Figure 7.40.

The FBG sensor is a type of nondestructive testing method used in situ on a structure to measure the vibrations generated by external excitation sources. When the vibration source consists of environmental loads (e.g., wind, traffic, or human activity), the structures are subjected to loads expressed as ambient oscillations or vibrations.

The ambient excitation force applied in this case study is numerically simulated using MATLAB (−0.4 to 0.4 N), as shown in Figure 7.41a. Its frequency range is initially detected at approximately 1–400 Hz, and its spectrum is displayed in Figure 7.4b. The ambient excitation force is perpendicular to the wall of the pipe. Three damage levels are introduced, that is, location: $D1$: 0.42–0.48 m, $D2$: 0.52–0.58 m, and $D3$: 0.62–0.68 m. It is assumed that all damage occurs toward the 180–360° range and is located on the internal surface of the pipe. Damage cases $D1$ and $D2$ have the same damage range but differ in location, while D3 has a greater damage range than the combined ranges of $D1$ and $D2$ ($D1 + D2$) [380, 427, 428].

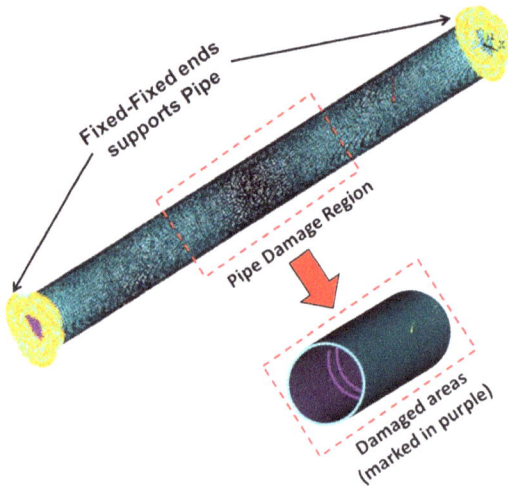

Figure 7.40: Fixed ends supporting the pipe.

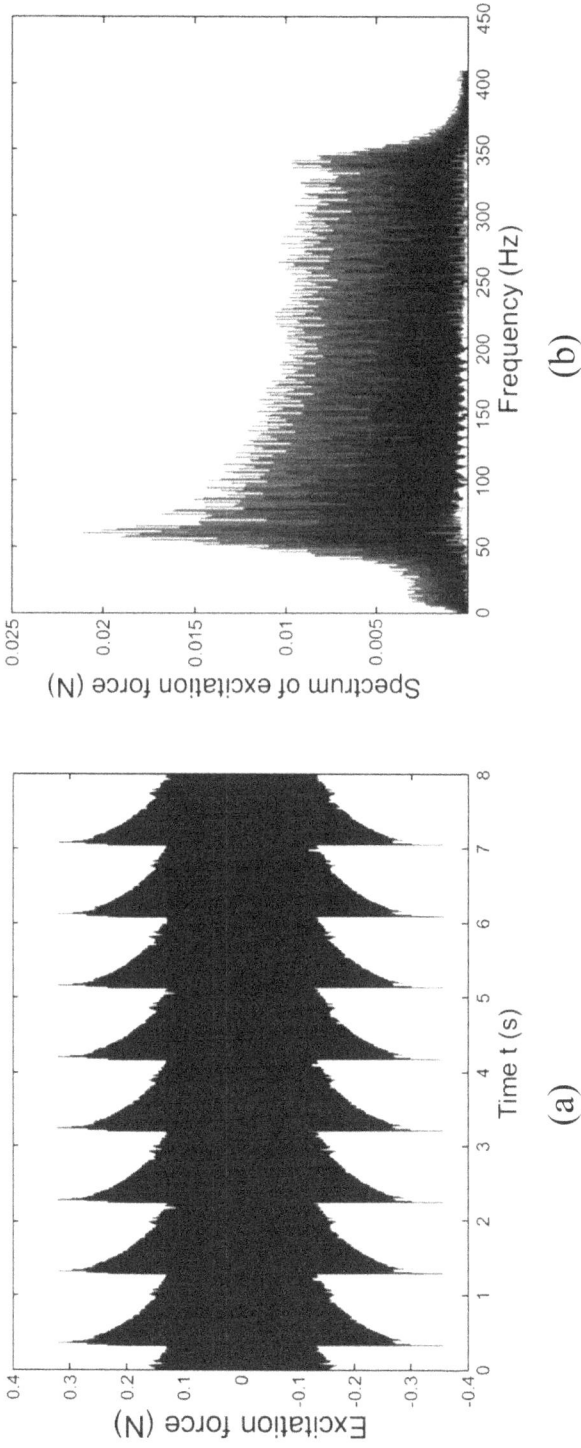

Figure 7.41: Ambient excitation forces: (a) time domain and (b) spectrum.

7.1.4.2.2 Modal analysis of the pipeline

The first, second, third, and fourth frequency mode shapes of the intact pipeline are shown in Figure 7.42 and Table 7.13 for various damage cases (D0–D3).

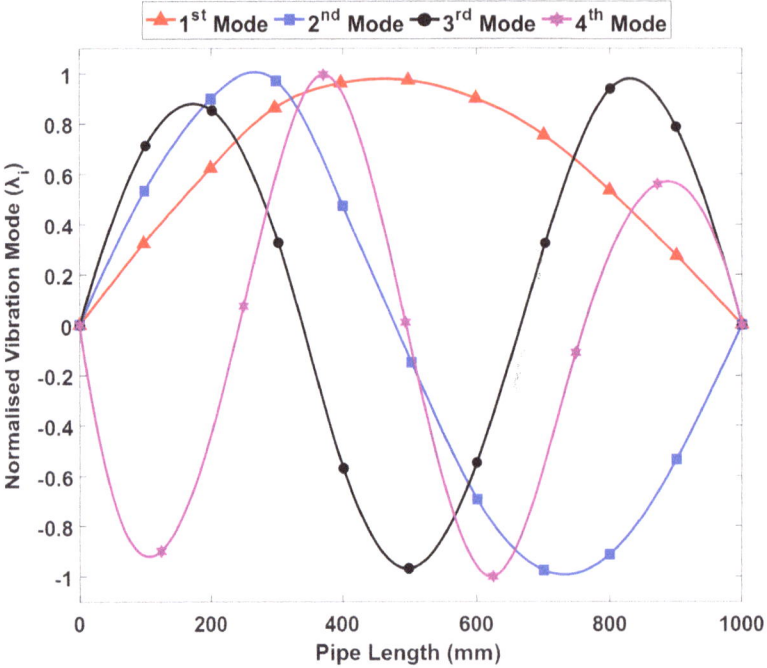

Figure 7.42: The first four mode shapes of an intact pipeline.

Table 7.13: Pipeline frequency rankings.

Damage	Frequency order			
	First order	Second order	Third order	Fourth order
D0 (Hz)	233.55	315.62	824.38	3,153.85
D1 (Hz)	232.85	312.67	822.90	3,151.45
D2 (Hz)	232.13	312.64	822.48	3,150.98
D3 (Hz)	230.79	310.11	820.85	3,147.73

Remark: D0 is UDP.

7.1.4.2.3 Stress–strain analysis in a stressed thick-walled pipe

Figure 7.43 shows a stressed thick-walled pipe with radius r. The solid line represents an unstrained pipe, while the dashed line represents a strained pipe.

As shown in Figure 7.43, the strain components of the pipe can be expressed as follows:

$$\varepsilon_H = \frac{2\pi(r+u)-2\pi r}{2\pi r} = \frac{u}{r}, \quad \varepsilon_L = \text{constant}, \quad \varepsilon_r = \frac{(\delta r + \delta u) - \delta r}{\delta r} = \frac{\delta u}{\delta r} \tag{7.20}$$

where ε_H is the hoop strain, ε_L is the longitudinal strain, and ε_r is the radial strain.

By applying the stress–strain relationship (tri-axial stresses) for a pipeline under hydrostatic internal pressure (**P**), the stress–strain relation can be expressed as follows:

$$\varepsilon_L = \sigma_L - \upsilon(\sigma_H + \sigma_r) = \sigma_L - \upsilon(\sigma_H - P) \tag{7.21}$$

$$E\varepsilon_H = E\frac{u}{r} = \sigma_H - \upsilon(\sigma_L + \sigma_r) = \sigma_H - \upsilon(\sigma_L - P) \tag{7.22}$$

$$E\varepsilon_r = E\frac{du}{dr} = \sigma_r - \upsilon(\sigma_H + \sigma_L) = -P - \upsilon(\sigma_H + \sigma_L) \tag{7.23}$$

where E is the elastic modulus, υ is Poisson's ratio, σ_H is hoop stress, σ_L is longitudinal stress, and σ_r is radial stress.

From the equilibrium of the pipe element shown in Figure 7.43, we can derive the differential equations for hoop stress as follows:

$$r(1-\upsilon)\frac{d\sigma_H}{dr} - r(1-\upsilon)\frac{dP}{dr} = 0 \tag{7.24}$$

By solving the above differential equations and applying the general boundary conditions of the pipe to the stress equation, the stress components in the pipe (see Figure 7.43) can be expressed as follows:

$$\sigma_H = \frac{Pr_i^2}{r_o^2 - r_i^2}\left(1 + \frac{r_o^2}{r^2}\right) \tag{7.25}$$

$$\sigma_H = \frac{Pr_i^2}{r_o^2 - r_i^2}\left(1 + \frac{r_o^2}{r^2}\right) \tag{7.26}$$

$$\sigma_L = \frac{Pr_i^2}{r_o^2 - r_i^2} \text{ for tube}, \quad \sigma_L = 0 \text{ for pipe} \tag{7.27}$$

The pipe deformation ($\delta = u$) in Figure 7.43 can be calculated using Equation (7.9):

$$\delta = \frac{(1-\upsilon)}{E}\left(\frac{Pr_i^2}{r_o^2 - r_i^2}\right)r + \frac{(1+\upsilon)}{E}\left(\frac{Pr_i^2 r_o^2}{r(r_o^2 - r_i^2)}\right) \tag{7.28}$$

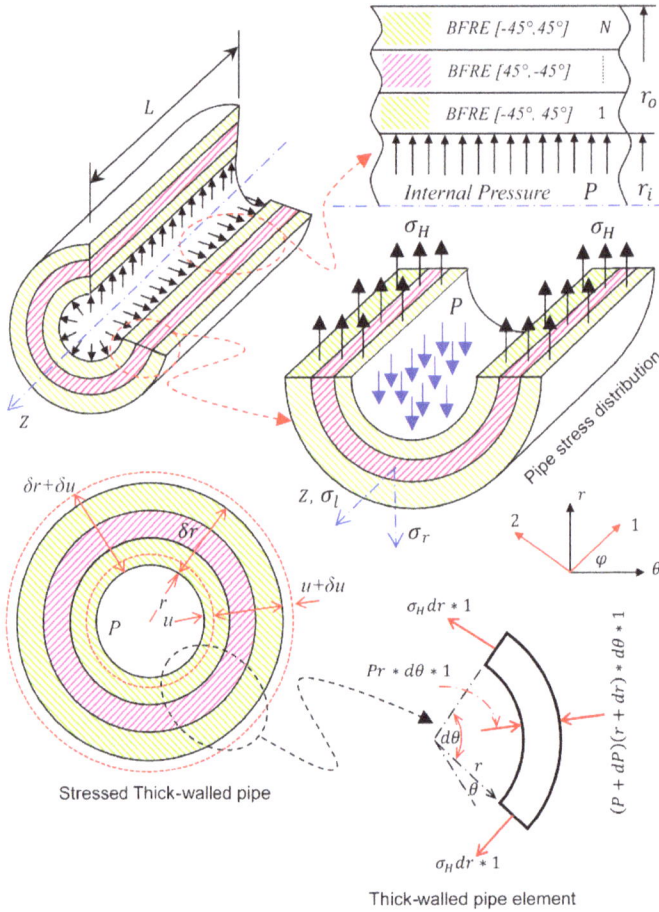

Figure 7.43: Stressed thick-walled pipe.

7.1.4.3 Pipeline monitoring-based fiber Bragg grating sensor technology

The FBG sensors that have been developed can be used to detect damage in pipelines. These include FBG-based pressure sensors for finding the point of leakage. The mechanisms of these sensors depend on the principle that when a leak initially occurs, there is a pressure drop in the pipeline on either direction of the leakage point. Determining the point of leakage depends on the time it takes for the wave to reach the FBG-based pressure sensors. The advantages and possible challenges of the FBG sensor system are as follows.

FBG sensors for pipelines can not only detect damage but also provide early warnings of hazardous pipeline events occurring along the pipeline. They possess features not found in traditional sensing technologies. The key FBG sensor capabilities are described as follows: The ability to measure the temperature and strain at thousands of points using a single fiber is highly significant for the inspection of slender structures (such as pipelines, oil wells, and coiled tubing).

(1) Reliable monitoring and detection of small and slow leakages of gas, oil, heating, and so on

(2) The ability to identify man-made pipeline damage and provide real-time alarms and positioning with a very high accuracy rate and a very low false alarm rate

(3) The system can provide an intelligent alarm function based on the geographic information system platform to meet monitoring requirements

(4) Continuous distributed measurement along the line without blind spots, featuring a self-diagnosis function, real-time detection, and localization of damage by the detection sensor system

(5) One optical cable can simultaneously perform temperature and strain detection and communicate at the same time

(6) High-temperature and low-temperature resistance, long-distance testing, and centralized ground signal processing

(7) Passive, intrinsically explosion-proof, especially suitable for use in flammable and explosive environments

(8) Good anticorrosion and anti-interference performance

(9) Long-term stability and measurement accuracy that are not affected by the loss of transmission fiber

7.1.4.4 Design theory of fiber Bragg grating strain sensor array

The sensor cannot gauge the hoop strain at a certain point until it is closely adhered to the outer wall surface of the pipeline. In this case study, due to the advantages of FBG distributed sensing, a single sensor with a multipoint monitoring system is used to monitor external strain changes. The FBG distributed sensor is installed on the outer wall surface of the pipeline in an arbitrary orientation (φ) as shown in Figure 7.44a. The relevant parameters of the packaged FBG are provided in Table 7.14.

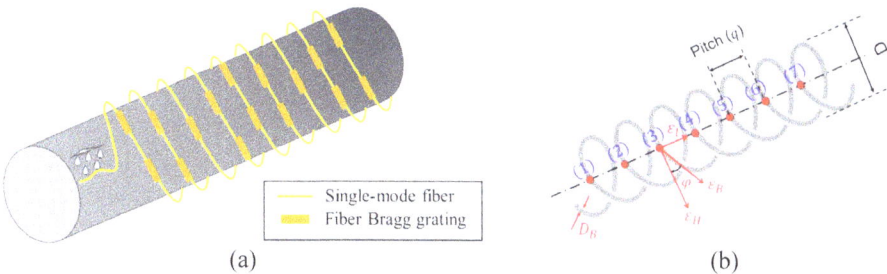

(a)   (b)

Figure 7.44: The theory of fiber Bragg grating strain sensor array: (a) fiber Bragg grating distributed sensors installed in the pipeline and (b) the relationship between the pipe hoop strain ε_H and the fiber grating strain ε_B.

Table 7.14: Fiber-optic parameters.

Parameter	Optical fiber
Elastic modulus (MPa)	$E_B = 70$
Diameter (mm)	$D_B = 0.125$
Area (mm^2)	$A_B = 0.01227$
Pitch (m)	$q_B = 0.1$
Length (m)	$L_B = 0.86\pi$
Arbitrary orientation	$\varphi = 45°$

From Figure 7.44b, the length of the FBG sensor (L_B) can be approximated as

$$L_B = \pi D N_t \tag{7.29}$$

where N_t is the total number of sensor coils, and D is the outer diameter of the sensor coils (see Figure 7.44b). N_t can be calculated using the following formula:

$$N_t = \frac{L - D_B}{q_B}, \quad D = 2r_o + D_B \tag{7.30}$$

where D_B is the sensor wire diameter and q_B is the pitch between the grating. Finally, the length of the FBG sensor can be estimated as

$$L_B = \pi \left[(2r_o + D_B) \frac{L - D_B}{q_B} \right] \tag{7.31}$$

The proposed sensor sensitivity can be determined by assuming that the relationship between the pipe hoop strain ε_H and the fiber grating strain ε_B is based on the premise that the hoop deformations of the sensor and pipe structures are the same, that is, the effect of sensor stiffness on the pipe is ignored. Moreover, the loss of strain due to the type of sensor fixation is not considered. The pipe hoop strain ε_H and the fiber grating strain ε_B can be expressed as follows:

$$\varepsilon_B = \frac{\Delta L_B}{L_B} = \frac{M}{E_B A_B} \tag{7.32}$$

where M is the internal force through the sensor caused by a certain change in the pipe's interior, and the hoop strain at the outer radius of the pipe ($r = r_o$) is given as

$$\varepsilon_H = \frac{\sigma_H}{E} = \frac{2Pr_i^2}{E(r_o^2 - r_i^2)} \tag{7.33}$$

And from Equation (7.29):

$$r_o = \frac{q_B}{2\pi} \left(\frac{L_B}{L} \right) \tag{7.34}$$

The relationship between the pipe hoop strain ε_H and the fiber grating strain ε_B, taking into consideration the angle of the sensor, φ as shown in Figure 7.44b, can be expressed as follows:

$$r_o = \frac{q_B}{2\pi} \left(\frac{L_B}{L \frac{\varepsilon_B}{\varepsilon_H} = \cos\varphi = \frac{ME}{2PE_B A_B} \left(\left(\frac{q_B}{2\pi r_i} \left(\frac{L_B}{L} \right) \right)^2 - 1 \right)} \right) \tag{7.35}$$

K_B is the strain sensitivity coefficient of the FBG strain sensor array and is given as

$$K_B = \frac{L_B}{L} = 0.86\pi \tag{7.36}$$

The relationship between the strain and the wavelength of the grating in this band can be approximated as follows:

$$\varepsilon_B = \frac{\Delta\lambda_B}{0.86\pi} \tag{7.37}$$

Therefore, the relationships between the center wavelength of the grating, the hoop strain ε_H, and the internal pressure P at the outer radius of the pipe ($r = r_o$) are determined by formulas (7.31), (7.34), and (7.35):

$$\varepsilon_H = \frac{\Delta\lambda_B}{0.86\pi \cos\varphi} \tag{7.38}$$

$$P = \frac{\Delta\lambda_B E \left(r_o{}^2 - r_i{}^2 \right)}{1.72\pi r_i{}^2 \cos\varphi} \tag{7.39}$$

Figure 7.45 illustrates the relationship between the wavelength of the various gratings of the sensor, the hoop strain ε_H, and the internal pressure P at the outer radius of the pipe ($r = r_o$).

As shown in Figure 7.45, the linear relationship between pressure, wavelength, and hoop strain, along with the sensor CF, reaches 0.9677, which indicates that there is little difference between the sensor gratings. From Figure 7.45, it can be concluded that the proposed sensor array has stable performance and is sensitive and suitable for monitoring the pipeline subject of this study.

7.1.4.5 The basalt fiber-reinforced polymer pipeline damage identification model

In Section 7.1.3.4, the sensitivity, stability, and linearity of the proposed FBG sensor network were verified for monitoring the pipeline, and the sensor response (pipeline displacement) was studied for both intact and damaged pipeline systems. However, identifying various levels of damage is challenging due to the inherent complexity of 3D modeling. Therefore, to effectively identify the damage, an efficient method must be used to extract the FBG sensor response features. In damage identification problems, a

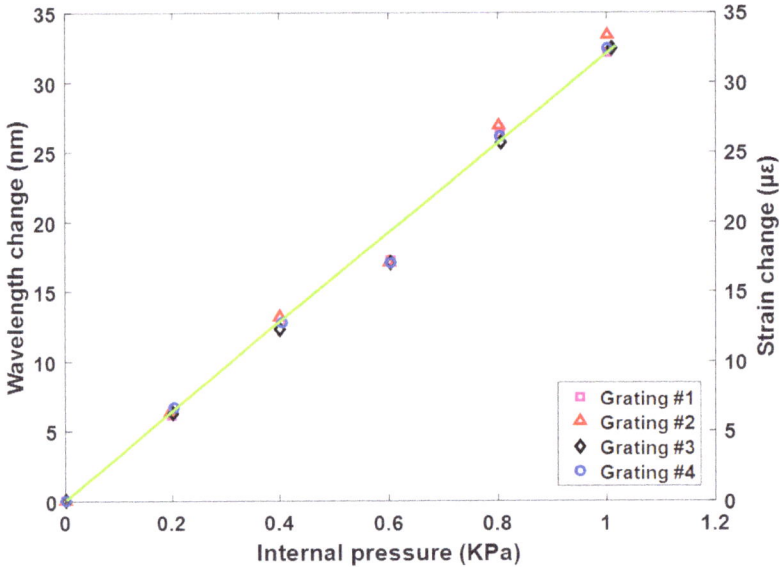

Figure 7.45: The relationship between the internal pressure of pipe and wavelength of the grating, and the hoop strain.

hybrid ECNN can be utilized to test the hypothesis presented in this research. This approach can yield better results than a traditional CNN + softmax (TCNN). To evaluate this, the classification response of a trained CNN is compared using neural codes learned from the same network, but with a k-nearest neighbor classifier applied to the output of the last hidden layer. Three configurations are evaluated to improve the accuracy of the CNN algorithm after training with standard backpropagation during the inference stage (see Figure 7.46):

(1) Using the CNN's softmax layer, referred to in this research as TCNN
(2) Using the last hidden CNN layer (before the softmax layer) to obtain the neural codes that are passed to a k-nearest neighbor, which compares them with the prototypes of the training set using the Euclidean distance to obtain the most likely class, known in this research as ECNN
(3) Using the k-nearest neighbor directly on raw data without any representation learning

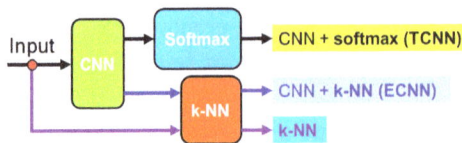

Figure 7.46: Scheme of the three configurations for improving accuracy.

7.1.4.5.1 A k-nearest neighbor algorithm

Because of its ease of implementation and high efficiency, the k-nearest neighbor algorithm is considered one of the most successful methods in past structural health monitoring applications. The application of the k-nearest neighbor method is based on searching for k points in a reference database for the closest points to the measured data, according to a function that represents the distance between them. This represents the optimal solution by minimizing the distance values of k:

(1) As the target of this new algorithm, the sensor displacement datasets and time–frequency spectrogram maps, both before and after damage, are utilized as training features for damage detection in composite pipe k-nearest neighbor damage prediction datasets with the following steps:

(2) The integer k should be assigned first.

(3) According to the dataset distribution, the optimized k is determined by iterating through integer values to identify the best k that yields the highest accuracy.

(4) The damage datasets undergo min–max normalization and z-score standardization, as shown in Equations (7.23) and (7.24).

This is because, when a pair of features is input, the k-nearest neighbor searches for the nearest k pairs of features using Euclidean distance on the same scale:

$$\text{Min–Max normalization}\ (X) = \frac{(X - \min(X))}{(\max(X) - \min(X))} \tag{7.40}$$

$$\text{z-score standardization}(X) = \frac{(X - \text{main}\ (X))}{\text{StdDev}\ (X)} \tag{7.41}$$

The performance of k-nearest neighbor results depends on the effectiveness of the method used to measure the distance between the model dataset features and new test inputs. The Euclidean method is the optimal method usually used for distance calculations between test and trained data. It measures the distance along a straight line from one point (x_1, y_1) to another (x_2, y_2):

$$\text{Euclidean distance} = \sqrt{\sum_{i=1}^{n} (x_i - y_i)^2} \tag{7.42}$$

Figure 7.47 illustrates the k-nearest neighbor algorithm. To determine the optimal number of neighbors (k), a genetic optimization method is used.

Figure 7.48 shows the flowchart of the proposed genetic algorithm for selecting the optimum k. At very small values of k ($k < 10$), the prediction error of the sensor displacement is very high, and the error decreases by increasing k from 10 to 100. The least error of prediction occurs at $k = 100$ and at $k > 100$, the error increases again. Table 7.15 presents the internal parameters of the k-nearest neighbor used in this study.

Table 7.15: Internal parameters of k-nearest neighbors.

Parameters	Value
Optimum neighbor number (k)	100
Optimization method	Genetic algorithm
Distance	Euclidean
Bucket size	50
Include ties	0
Distance weight	Equal
Break ties	Smallest
Standardize data	1
Type	Prediction
min(X)	[2.840, 41.322]
Std Dev (X)	[0.485, 1.016]
Weight (W)	74.9×10^{-5}

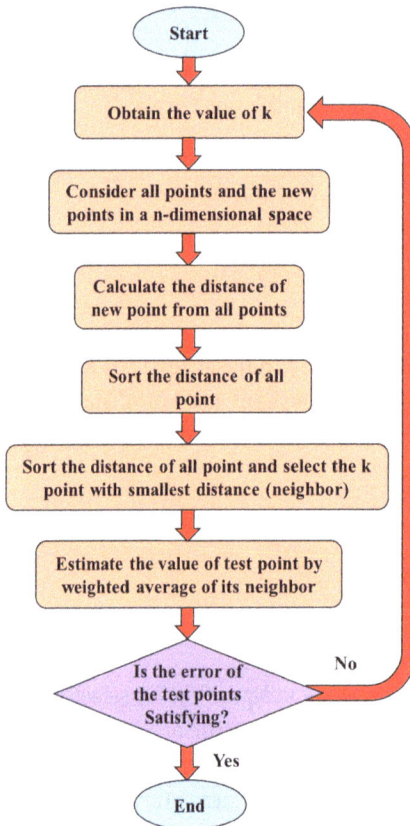

Figure 7.47: Flowchart of the k-nearest neighbors algorithm.

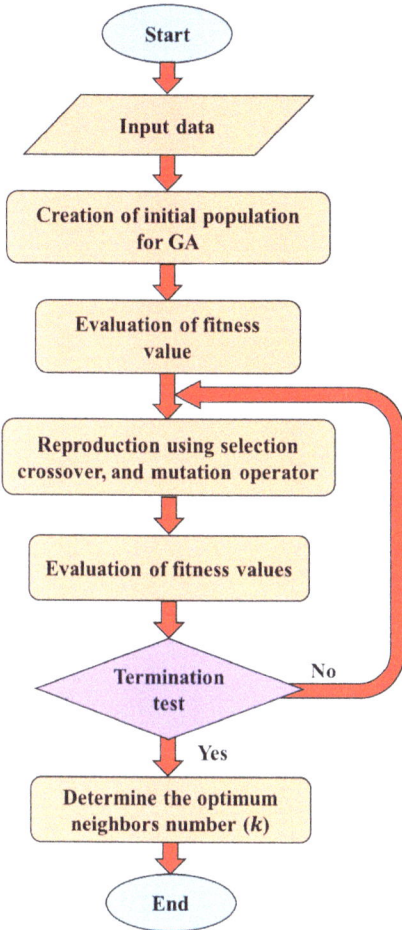

Figure 7.48: Flowchart of the proposed genetic algorithm for selecting the optimal k.

7.1.4.5.2 The convolutional neural network modeling

The architectural design of the proposed CNN layer is illustrated in Figure 7.49. As depicted in Figure 7.49, the CNN structure typically features a multilayer architecture, comprising convolution, pooling, activation, and fully connected layers. The network's input consists of time–frequency spectrogram images of the pipeline, captured both before and after damage. Significant local features are subsequently extracted through the convolutional and pooling layers. Finally, damage identification is performed via the fully connected layer.

The fully connected layers in the CNN architecture are located, as shown in Figure 7.50, between the input (features) and the output (target prediction) layers. These layers facilitate the extraction of higher-level features during the training process.

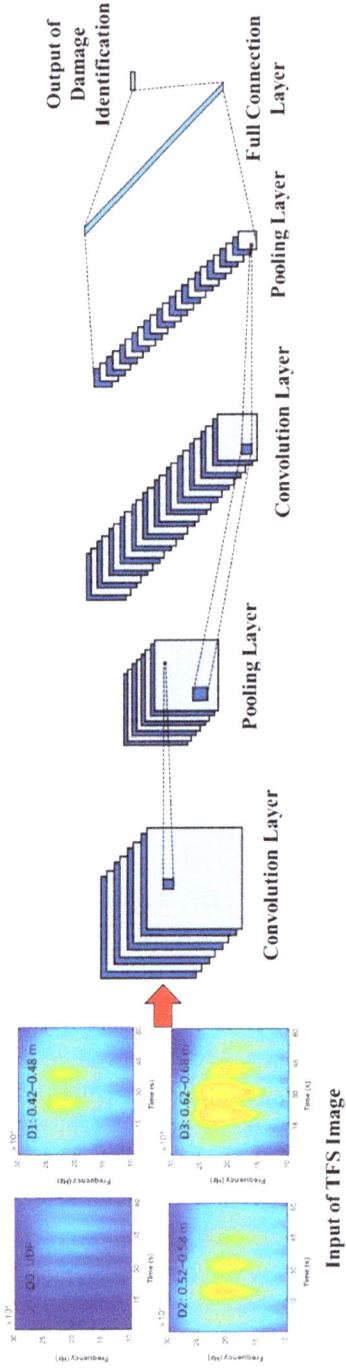

Figure 7.49: The architecture of a typical convolutional neural network.

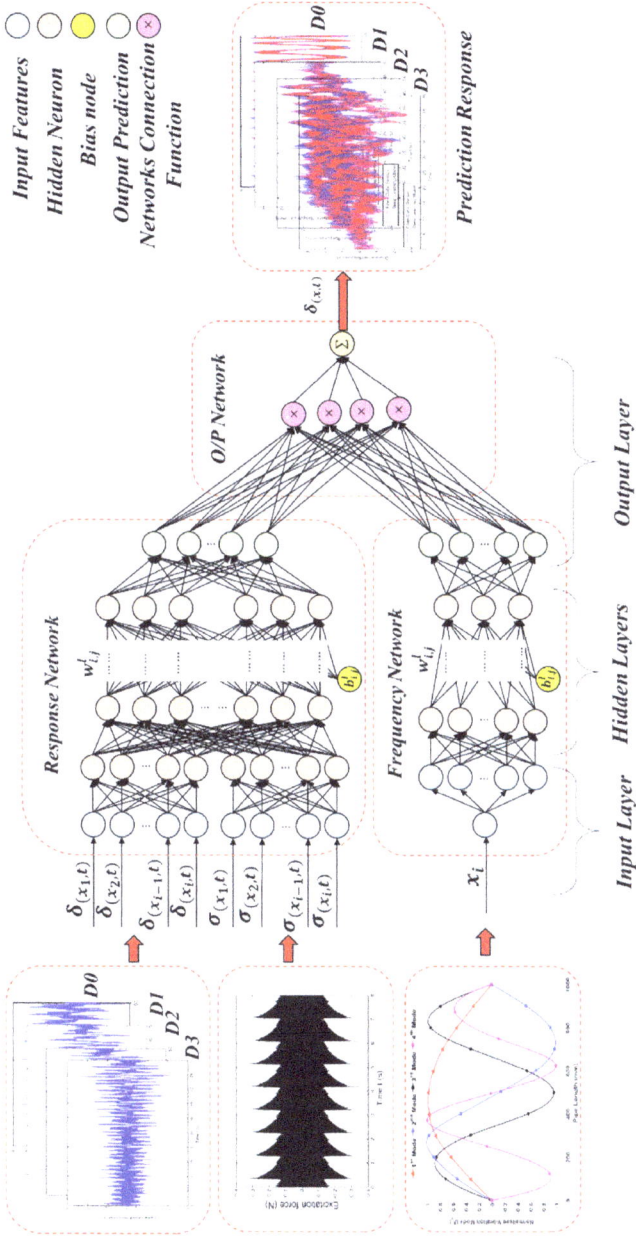

Figure 7.50: The architecture of the proposed ECNN for response prediction.

Using weights w and biases b, the CNN is trained to identify more useful nonlinear information. In the proposed CNN, this process is defined by Equation (7.2).

One important component of a neural network is the activation function, which provides the necessary nonlinearity, where a neural network becomes simple linear without it (essentially a probability distribution).

Softmax is used for multi-classification problems, where it normalizes the neural network's output to fit between "0" and "1." It is applied to represent the certainty, "probability" in the network's output. The expression for the softmax activation function is given as follows:

$$\sigma(\vec{z})_i = \frac{e^{z_i}}{\sum_{j=1}^{K} e^{z_i}} \tag{7.43}$$

where $\vec{z}$ is the input vector, z_i values are the elements of the input vector, e^{z_i} is the standard exponential function applied to each element of the input vector, and K is the number of classes in the multi-class classifier.

The pooling layer is usually arranged between sequential convolution layers. It is used to reduce the spatial dimensions of feature maps. This process is also referred to as undersampling, which helps control network overfitting. The operations commonly used for undersampling are maximum pooling and average pooling. The average pooling functionality of the pooling layer can be represented mathematically in Equation (7.44), assuming the pooling size is c, jth is the region, and lth is the number of pooling layers:

$$x_j^l = f\left(B_j^l \operatorname{mean}\left(x_j^{l-1}\right) + b_j^l\right) \tag{7.44}$$

where B_j^l is multiplicative and mean$(\cdot)$ is the average operation. The convolution and pooling layers work together to detect local connections, merge similar features, and remove unnecessary or irrelevant details.

7.1.4.6 The displacement response identification

Figure 7.51 shows the displacement caused by ambient excitation, as measured by the FBG sensor, for the undamaged pipe (UDP) model $D0$ and the damaged pipe (DP) models $D1$, $D2$, and $D3$, respectively. The boundaries are fixed for both ends of the pipeline. At a distance of 0.2 m from the pipeline, right support is the position at which the ambient forces are applied. The various levels of damage along the pipeline are taken into account.

Figure 7.52 shows the time–frequency spectrogram for the UDP model $D0$ and the DP models $D1$, $D2$, and $D3$, respectively. The time–frequency spectrograms in this case study are extracted from the direct wave packet before and after pipe damage, with the frequency of the Gaussian wavelet transform set between 50 Hz and 300 Hz at intervals of 1 Hz. As shown in Figure 7.52, the time–frequency spectrogram images

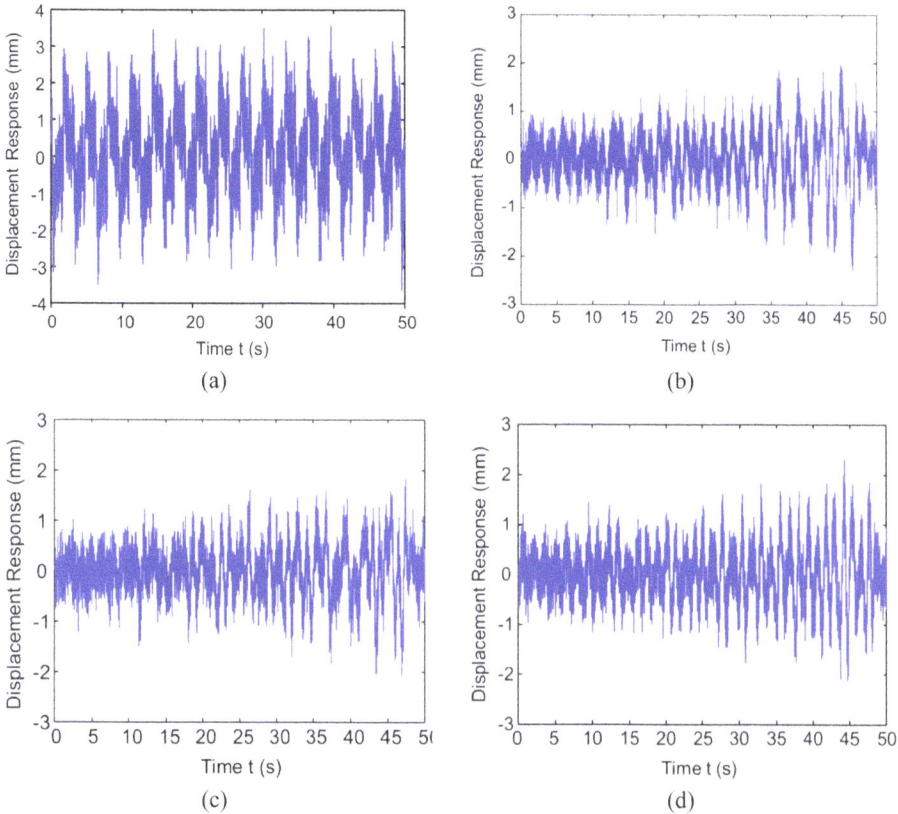

Figure 7.51: The displacement response of the UDP/DP: (a) UDP, $D0$; (b) DP, $D1$; (c) DP, $D2$; and (d) DP, $D3$.

have dimensions of 201 × 213. In Figure 7.52a–d, the magnitude of the direct wave packet increases because the reflected wave, caused by pipe damage, becomes more pronounced as the extent of damage changes.

7.1.4.7 Experimental validation of the proposed method

In this section, we present experimental validations of the proposed approach by applying the technique to a dataset adapted from Wang et al. [429]. They used an optical fiber sensing system configured with FBG to extract the damage behavior in carbon fiber-reinforced polymer composite pipes. The ultimate goal of this section is to extract the experimental dataset from Wang et al. [429] and compare it with the numerical dataset presented in this case study, which is derived from a sensory network of FBG series installed on the outer surface of the pipes. The aim is to prove the effectiveness and feasibility of the proposed sensing technique.

The vibration experiments were performed on the cantilevered carbon fiber-reinforced polymer pipes shown in Figure 7.53. Excitation frequencies of 10 Hz were

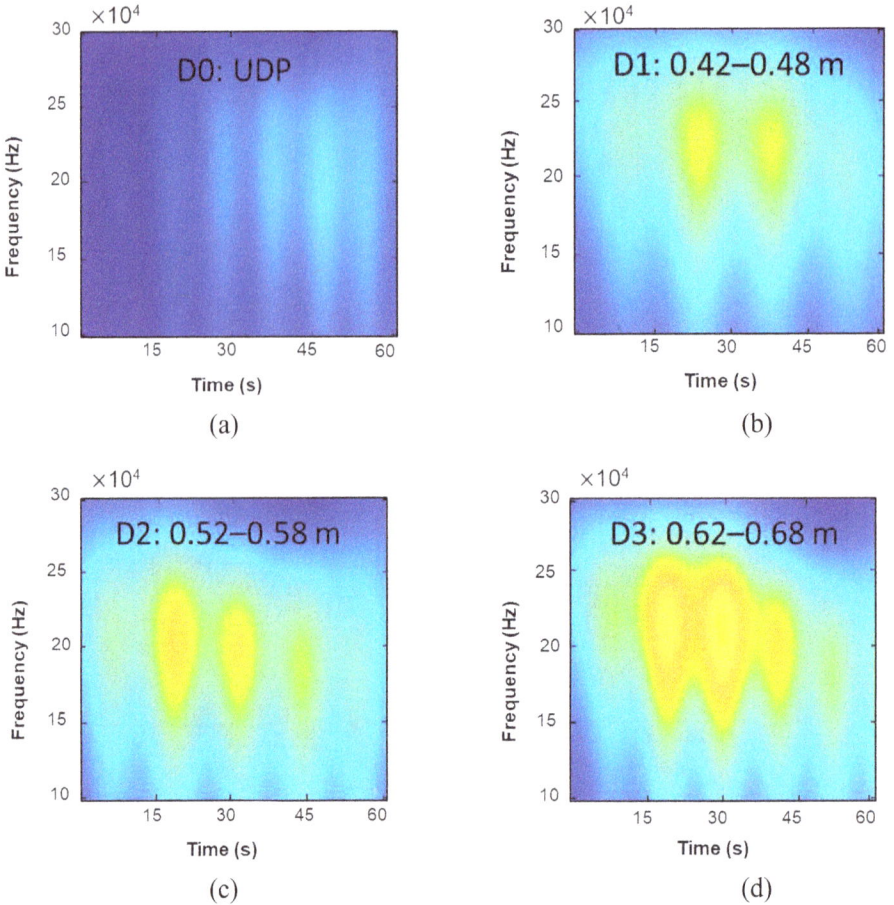

Figure 7.52: The time–frequency spectrogram of the UDP/DP: (a) UDP, *D0*; (b) DP, *D1*; (c) DP, *D2*; and (d) DP, *D3*.

applied to the specimens with a loading of about 2,000 s. The dimensions of the specimens were 12 mm, 15 mm, and 100 mm for the inner diameter, outer diameter, and length, respectively. Three series of FBGs were installed on the outer surface of the pipe as follows: the first two series were composed of three FBGs (FBG5, 6, 7), and the last series had four FBGs (FBG1, 2, 3, 4).

The vibration strain values at $x = 0.75$ m and adjacent to the FBG points were extracted numerically from the finite element model and experimentally from the FBG and then compared. Figure 7.54 presents the distribution of the vibration strain for both the computed and measured values over time.

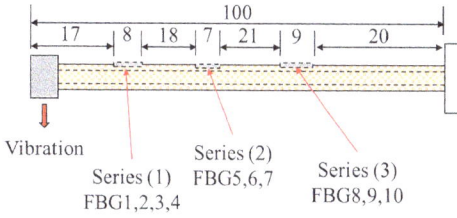

Figure 7.53: The positions of the fiber Bragg gratings in the series setup of the pipe.

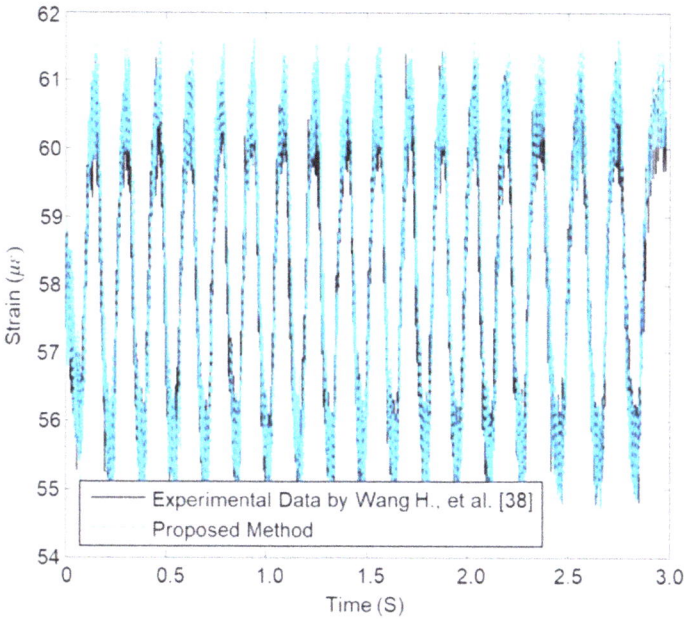

Figure 7.54: Comparison between the proposed method and the experimental values of strain at 10 Hz for FBG2 in a carbon fiber-reinforced polymer composite pipe.

7.1.4.8 Hybrid convolutional neural network + k-nearest neighbor (ECNN) architecture as a surrogate model

In the final layer of the CNN, the learned feature maps are flattened into a single vector by a fully connected layer, and the expected output is extracted. In this case study, the displacement response evaluation problem for UDP and DP is a regression problem. Therefore, either a softmax activation function or a k-nearest neighbor algorithm is adopted in the fully connected layer, by which a vector value that represents the displacement response is outputted. As discussed in Section 7.1.3.5, the proposed model (ECNN) architecture consists of two convolutional layers, two subsampling layers, and a fully connected layer. Each layer adjusts the parameters and corresponding weights. The overall process of the proposed ECNN is illustrated in Figure 7.55.

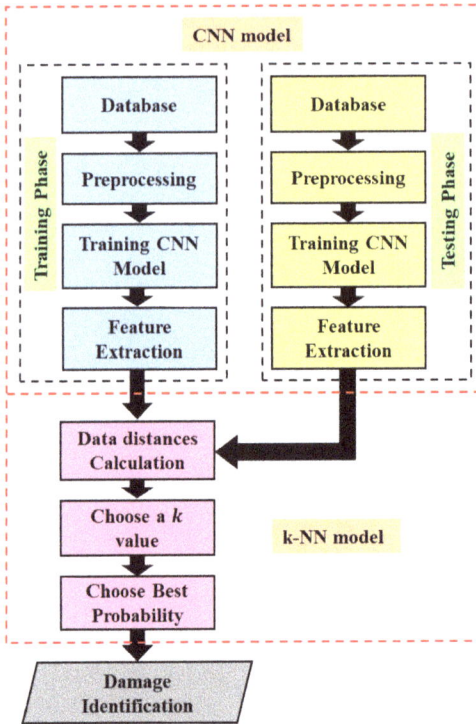

Figure 7.55: The overall convolutional neural network + k-nearest neighbor (ECNN) flowchart.

The prediction of damaged pipeline displacement can be described through the steps outlined in Figure 7.56. If the ambient excitation $\sigma(x,t)$ and the initial conditions are provided, the proposed ECNN can be utilized to predict the pipeline's displacement response.

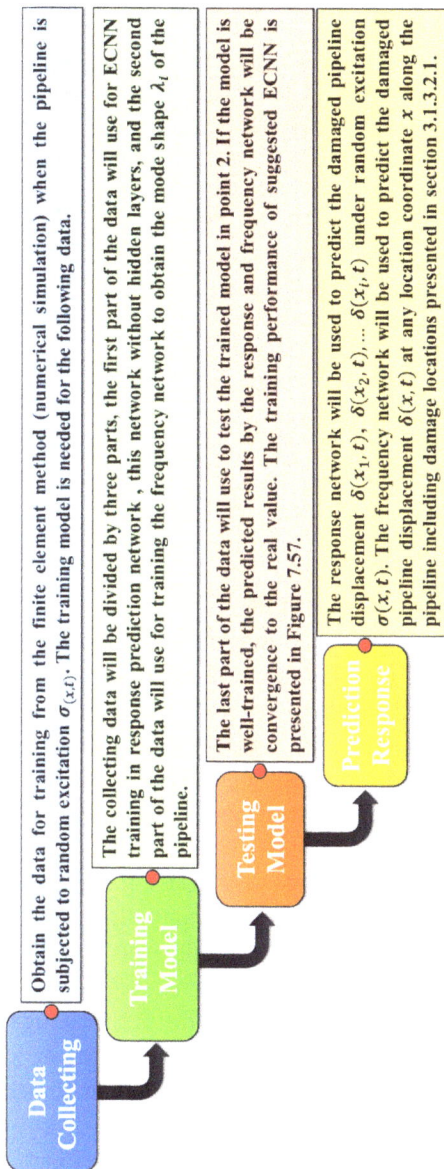

Data Collecting

Obtain the data for training from the finite element method (numerical simulation) when the pipeline is subjected to random excitation $\sigma_{(x,t)}$. The training model is needed for the following data.

Training Model

The collecting data will be divided by three parts, the first part of the data will use for ECNN training in response prediction network , this network without hidden layers, and the second part of the data will use for training the frequency network to obtain the mode shape λ_i of the pipeline.

Testing Model

The last part of the data will use to test the trained model in point 2. If the model is well-trained, the predicted results by the response and frequency network will be convergence to the real value. The training performance of suggested ECNN is presented in Figure 7.57.

Prediction Response

The response network will be used to predict the damaged pipeline displacement $\delta(x_1, t)$, $\delta(x_2, t)$, ... $\delta(x_i, t)$ under random excitation $\sigma(x, t)$. The frequency network will be used to predict the damaged pipeline displacement $\delta(x, t)$ at any location coordinate x along the pipeline including damage locations presented in section 3.1.3.2.1.

Figure 7.56: The steps of ECNN development.

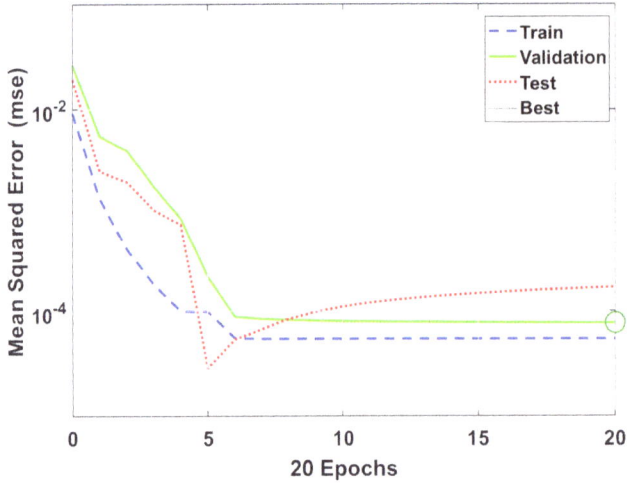

Figure 7.57: Training performance of the proposed ECNN.

7.1.4.9 The displacement response prediction based on ECNN

As shown in Figure 7.41, an ambient excitation load $\sigma(x,t)$ in a composite pipeline is used to obtain the training data. The trained model is then utilized to predict the displacement response of the damaged pipeline under the same excitation load. Figure 7.58 illustrates the training data. As demonstrated in the figure, the proposed deep learning model can predict the displacement response of the composite pipeline under ambient excitation load. The results indicate excellent agreement between the FBG sensor data and the ECNN data, with an average error of 0.093%.

Table 7.16 lists all the cases related to various damage cases (D0–D3). Table 7.17 presents a comparison between the three artificial intelligence configurations for damage identification results for four labels, based on TPR, TNR, FPR, and FNR.

In general, for all indices (TPR, TNR, FPR, and FNR), using k-nearest neighbor over the input datasets resulted in a lower average accuracy than the TCNN configuration. The hybrid approach, ECNN, achieved better results than the TCNN or k-nearest neighbor. Overall, the proposed ECNN approach consistently outperformed the TCNN and k-nearest neighbor across all indices.

To estimate the proposed ECNN's performance in damage identification of basalt fiber-reinforced polymer composite pipelines, three indicators were calculated during the training process. The accuracy rate (P%), regression rate (R%), and F1-score (F%) are based on the indices (TPR, TNR, FPR, and FNR) and are calculated as follows:

$$P\% = \frac{\text{TPR}}{\text{TPR} + \text{FPR}} \tag{7.45}$$

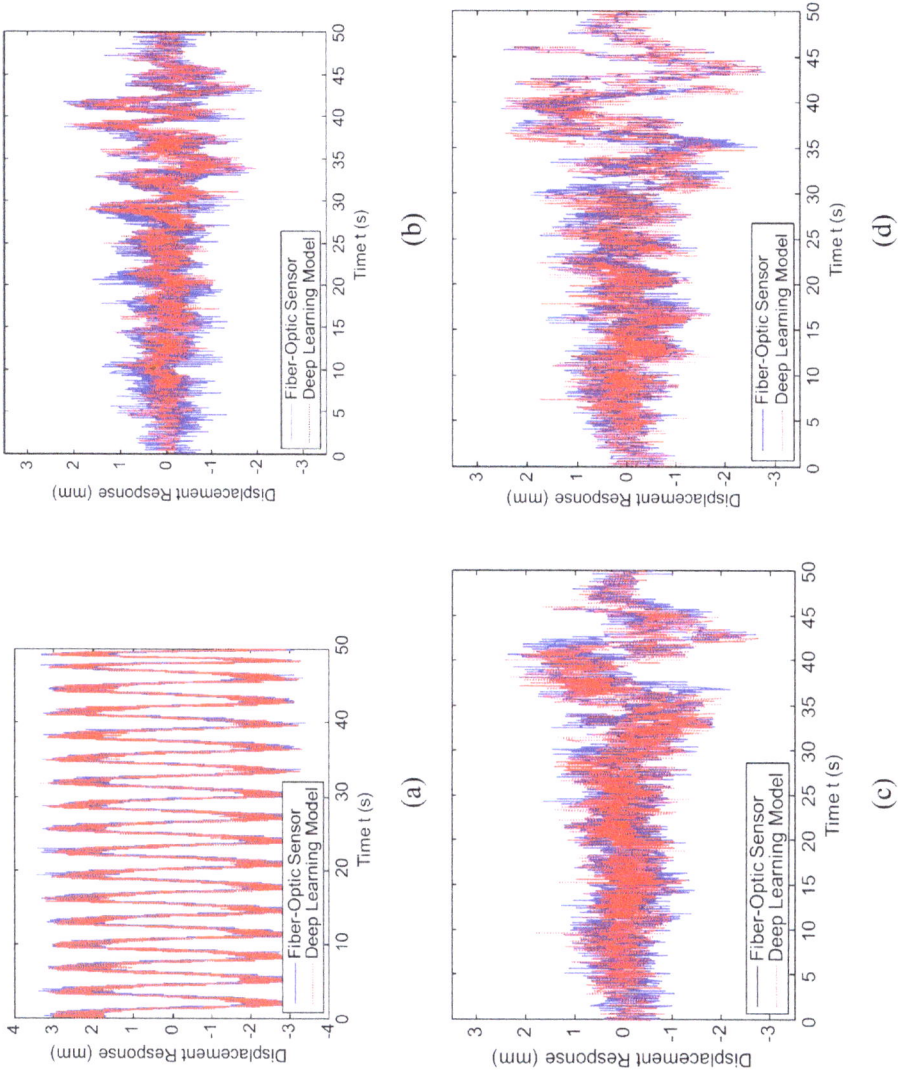

Figure 7.58: The displacement response prediction of the UDP and DP under ambient excitation load $\sigma(x,t)$: (a) response prediction of UDP, $D0$; (b) response prediction of DP, $D1$; (c) response prediction of DP, $D2$; and (d) response prediction of DP, $D3$.

$$R\% = \frac{\text{TPR}}{\text{TPR} + \text{FNR}} \qquad (7.46)$$

$$F\% = \frac{2\,\text{TPR}}{2\,\text{TPR} + \text{FNR} + \text{FPR}} \qquad (7.47)$$

Figure 7.59 shows the calculated $P\%$, $R\%$, and $F\%$ for 800 testing datasets, which are divided into 10 groups of displacement results (i.e., $D1$, $D2$, . . ., and $D10$) after 350 iterations to verify the effectiveness of the proposed ECNN. These results are presented as bar plots in different colors, and the overall performance of the selected displacement is indicated with a dashed line.

As shown in Figure 7.59, the overall performance values were 93.33%, 91.18%, and 90.54%, for $P\%$, $R\%$, and $F\%$, respectively. These results confirm that the proposed method can automatically identify damage in composite pipelines with satisfactory performance, regardless of the corresponding capacitance data noise background and conditions.

As shown in Figure 7.59, based on the results of testing with 800 datasets, the proposed method, which applies ECNN to identify varying degrees of damage in pipelines, is promising and may also be suitable for other composite structures.

Table 7.16: Labeled datasets.

Label	1	2	3	4
Case	D0	D1	D2	D3
Location	UDP	0.42–0.48 m	0.52–0.58 m	0.62–0.68 m

7.1.4.10 Summary

Based on the response of the FBG sensor system in the damaged composite pipeline, a novel artificial intelligence-based algorithm is proposed. This algorithm combines deep learning and machine learning by utilizing an ECNN without modifying the training stage for the displacement response prediction of the composite pipeline. The proposed architecture replaces the softmax layer in TCNN with a k-nearest neighbor algorithm for inference. The proposed ECNN model is divided into two networks: a response network and a frequency network. The frequency network converges to the shape mode of the pipeline, while the response network serves as a feedback mechanism for predicting the displacement response of a long pipeline. A composite pipeline made of basalt fiber-reinforced polymer was analyzed using a finite element model to simulate the pipeline's displacement. Three damage levels were introduced to validate the effectiveness of the proposed approach. The training data were generated using the finite element model. From the results, it can be concluded that the proposed artificial intelligence-based model can effectively predict the displacement response of composite pipelines and works much faster in terms of computational time than the traditional

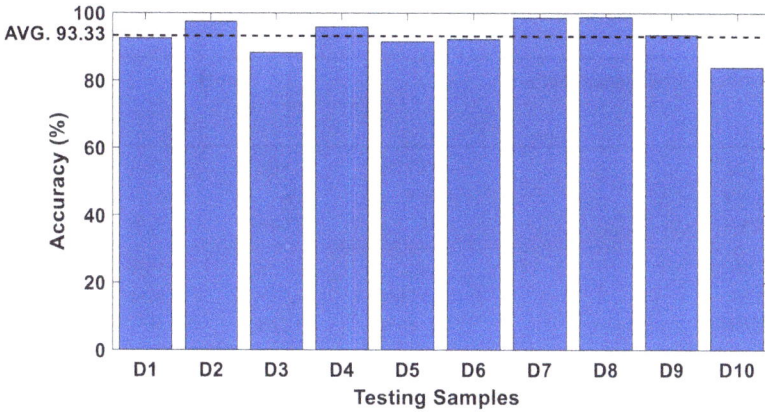

(a)

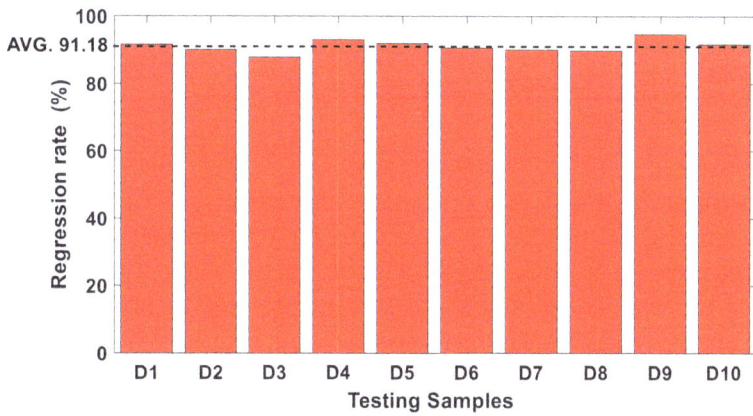

(b)

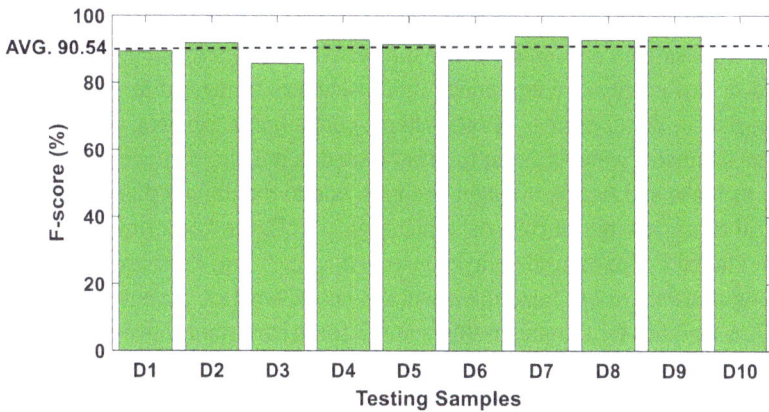

(c)

Figure 7.59: Comparison of the training process based on damage identification test data for 10 datasets of displacement: (a) accuracy (P%); (b) regression rate (R%); and (c) F1-score (F%).

Table 7.17: Identification of testing detailed results.

Indices	Artificial intelligence method	Label			
		1	2	3	4
TPR	k-NN	97.11%	90.89%	91.73%	93.97%
	TCNN	98%	92%	93%	95.2%
	ECNN	98.71%	95.64%	96.32%	97.19%
TNR	k-NN	94%	96.82%	97.13%	95.81%
	TCNN	95.66%	98.53%	97.95%	96.74%
	ECNN	97.61%	99.32%	99.46%	98.72%
FPR	k-NN	6.21%	2.14%	4.34%	6.65%
	TCNN	6.35%	2.45%	4.56%	6.99%
	ECNN	7.59%	3.49%	6.1%	8.32%
FNR	k-NN	3.87%	11.98%	11.27%	10.34%
	TCNN	4%	13%	12%	11.3%
	ECNN	6.21%	15.4%	13.64%	12.33%

k-NN is k-nearest neighbor algorithm.

finite element model. The results also demonstrate that replacing the softmax layer with the k-nearest neighbor significantly outperforms the TCNN architecture in terms of accuracy. The proposed method achieved satisfactory performance, with the values of $P\%$, $R\%$, and $F\%$ being 93.33%, 91.18% and 90.54, respectively. Therefore, the proposed method offers distinct advantages for solving practical engineering problems.

7.2 Structural health monitoring of composite ducts

A muffler/duct is considered a common device for passive noise elimination. This device is widely used in various types of equipment and machinery connected to ducts or exhaust systems, such as diesel engines, HVAC systems, compressors, blowers, and fans for ventilation. The laminated composite muffler is also used in military equipment and machinery, such as tanks and heavy and light cannons, due to the unique thermoacoustic and mechanical characteristics of composite materials (see Figure 7.60). However, mufflers are often limited by space availability. Consequently, there has been growing interest in designing mufflers to optimize their acoustic characteristics. Several optimized transmission loss methods for shaping mufflers under space constraints have been developed by several researchers over the last two decades. Numerous studies have also been carried out by many researchers on mufflers to estimate the noise attenuation performance. These studies have suggested various evaluation methods for the actual noise reduction in mufflers when an optimally designed muffler is mounted on a duct.

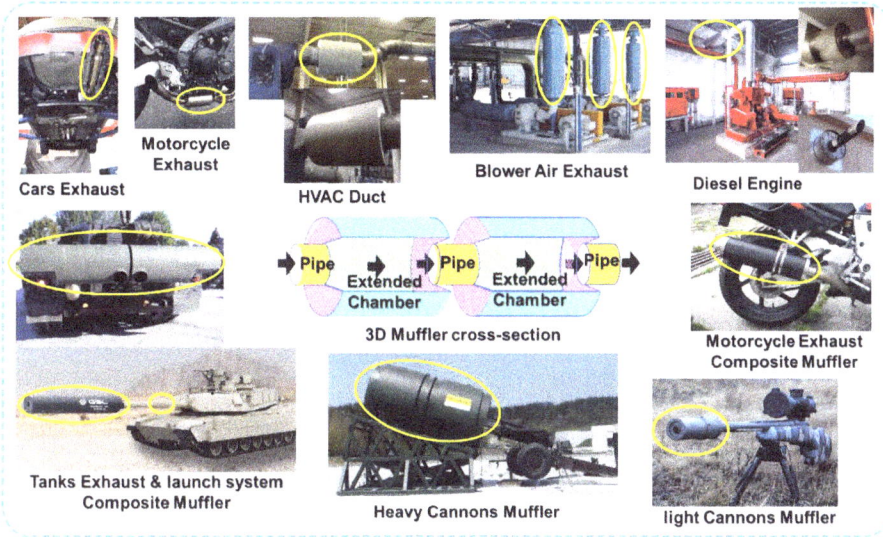

Figure 7.60: Examples of industrial and military equipment that are connected with mufflers.

7.2.1 Case Study (4): structural health monitoring of composite dual-chamber muffler using deep learning algorithm for predicting acoustic behavior

Over the last two decades, several experimental and numerical studies have been conducted to investigate the acoustic behavior of various muffler materials. However, there is a problem when it becomes necessary to perform large, important, time-consuming calculations, particularly if the muffler is made from advanced materials such as composite materials. Therefore, this case study introduces the concept of an indirect dual-chamber muffler made from a basalt fiber-reinforced polymer laminated composite. It incorporates a monitoring system that utilizes a deep learning algorithm to predict the acoustic behavior of the muffler material, thereby saving time and effort in muffler design optimization. Two types of deep neural network architectures are developed in Python. The first deep neural network is a recurrent neural network with long short-term memory (RNN-LSTM) blocks, while the second is a CNN. First, a dual-chamber laminated composite muffler (DCLCM) model is developed in MATLAB to provide acoustic behavior datasets of mufflers such as acoustic transmission loss and the power transmission coefficient. The model's training parameters are optimized using Bayesian genetic algorithm optimization. The acoustic results from the proposed method are compared with experimental results available in the literature, thereby validating the accuracy and reliability of the proposed technique. The results indicate that the present approach is efficient and significantly reduces the time and effort to select the muffler material and optimal design, where both CNN and RNN-LSTM models achieved accuracy levels above 90% on the test and validation datasets. This work will reinforce the muffler industry, and its design may one day be equipped with deep learning-based algorithms.

7.2.1.1 Methodology

The presented case study utilizes two types of artificial neural network architectures: RNN-LSTM and CNN. These two types of artificial neural networks were developed in Python to predict the acoustic behavior of a DCLCM manufactured from a basalt fiber-reinforced polymer laminated composite. Figure 7.61 illustrates a flowchart of the proposed method. The power transmission coefficient and the transmission loss of an acoustic muffler are calculated using the exact solution of the governing acoustic equations for the muffler model. Once the essential characteristics of the muffler are extracted, these characteristics are fed into an artificial neural network as labeled data. The input and output weights of the network are then calculated. Subsequently, the derivative of the error for the weights is computed using the backpropagation algorithm. If the error target is not acceptable "case no," the muffler characteristics data are fed back into the artificial neural network to retrain the model and calculate new weights, thereby computing a new error. This process is repeated until the error target becomes acceptable "case yes." The training and model parameters were optimized using Bayesian genetic algorithm optimization. The proposed case study will be discussed in greater detail in the following sections.

In this case study, a novel deep learning algorithm is utilized to develop a monitoring strategy for the acoustic behavior performance of DCLCM. The muffler's acoustic performance is calculated for a wide range of geometric data of the DCLCM using commercial software. An RNN-LSTM and a CNN are trained, evaluated, and developed using Python. A Bayesian genetic algorithm optimization is used to optimize the training and parameter selection for the model. To the best of the authors' knowledge, the methodology presented in this case study and the results obtained represent an original contribution to scientific research in monitoring the acoustic behavior of composite muffler materials.

7.2.1.2 Materials and methods

7.2.1.2.1 The geometric model

Figure 7.62 shows the geometry of a circular DCLCM considered in this study. An acoustic DCLCM was constructed using a basalt fiber-reinforced polymer laminated composite, the staking distribution to three symmetrical plies is $[0/90°/0]_s$, and the thickness of each ply is 5 mm. The physical and mechanical properties of the basalt fiber-reinforced polymer are presented in Table 7.18. The geometric information used for the calculations in this case study is summarized in Table 7.19. The values of the muffler thickness, B, to the total chamber length, L, are listed in Table 7.19. The frequency range used for calculating the power transmission coefficients of the muffler

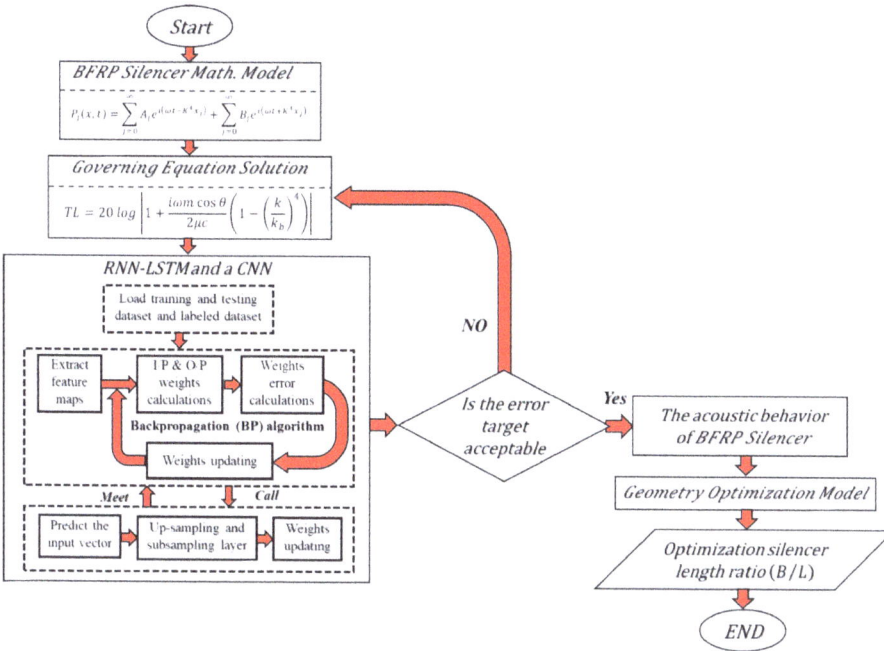

Figure 7.61: Flowchart of the proposed method.

is 0–3 kHz. The muffler length, B, is optimized to maximize the acoustic transmission loss. At the right end of the muffler, an anechoic termination is assumed.

Table 7.18 presents the physical and mechanical properties of the basalt fiber-reinforced polymer, where ρ is the material density; E_{11} and E_{22} denote the elastic modulus in the "1" and "2" directions, respectively; G_{12} and G_{21} represent the shear modulus; and v_{21} and v_{12} indicate the Poisson's ratio of the transverse strain in the directions "1" and "2" caused by the normal stress in the directions "2" and "1," respectively.

7.2.1.2.2 Basic acoustic equations of the dual-chamber muffler
Figure 7.63 presents the mathematical model of DCLCM which consists of three straight pipes and two expansion chamber pipes being identified. As shown in the figure, $S_1 = S_3 = S_5$ is the area of the straight pipe, and $S_2 = S_4$ is the area of the muffler. Eight points were chosen to represent the flow condition inside the muffler (pt1 ~ pt8). At each two consecutive points, the standing pressure P_s and reflected P_r pressure are similar, for example, $P_{s_1} + P_{r_1} = P_{s_2} + P_{r_2}$; in addition, the standing U_s volume velocity and reflected U_r volume velocity are similar, for example, $U_{s_1} + U_{r_1} = U_{s_2} + U_{r_2}$. The continuity algorithm of pressure and volume velocity on continuity junction numbers $(1-1)$, $(1-2)$, $(2-1)$, $(2-2)$ are applied to compute junctions' transmission coeffi-

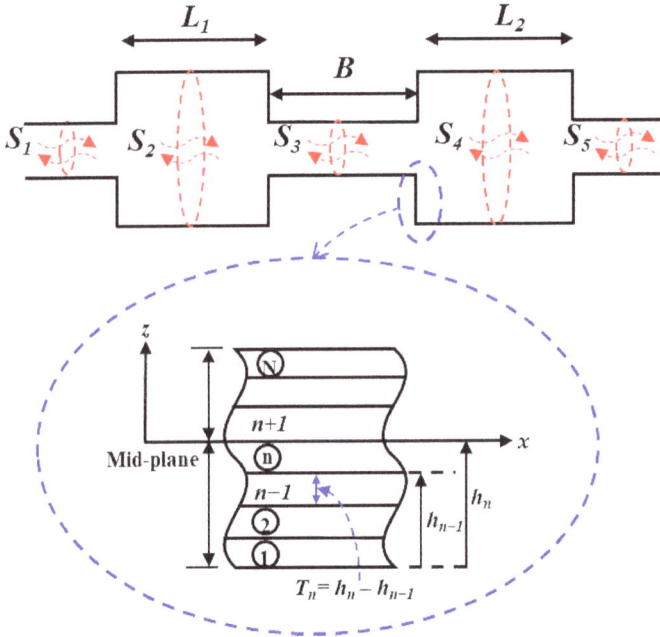

Figure 7.62: The geometrical model of the DCLCM.

Table 7.18: Physical and mechanical properties of basalt fiber-reinforced polymer.

E_{11} (GPa)	E_{22} (GPa)	G_{12} (GPa)	G_{21} (GPa)	$\upsilon_{12\,v1}$	υ_{21}	$\rho(kg/m^3)$
96.74	22.55	10.64	8.73	0.3	0.6	2,700

Note: This mechanical parameter or property can be found in [430].

Table 7.19: The parameters' specific values of the DCLCM model used for acoustic analysis.

Symbol	Description	Value (unit)						
B/L	A muffler to chamber length (S/C) ratio	0	0.1	0.2	0.4	0.6	0.8	1.0
R_C	Chamber radius	0.1 (m)						
R_P	Inlet pipe radius	0.05 (m)						
L	Total chamber length	0.2 (m)						

cients A_{11}, B_{11}, A_{12}, B_{12}, A_{21}, B_{21}, A_{22}, B_{22}, by solving the Helmholtz equation (7.48) [30], and applying the junction boundary condition at each junction:

$$\nabla^2 P + k^2 P = 0 \tag{7.48}$$

For the equal-sized chamber systems shown in Figure 7.63, the three-point method was used to describe the acoustical properties via absorbing materials, assuming the notation of plane wave propagation, as the transverse system dimensions are smaller than one wavelength over the frequency range of interest.

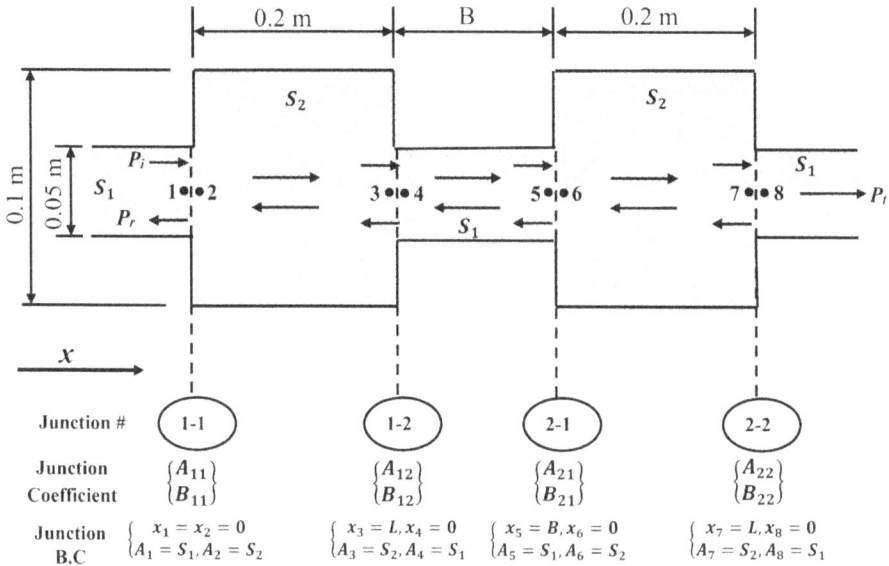

Figure 7.63: The mathematical space constraints for DCLCM.

A standing pressure propagating wave amplitude A_j and a reflected pressure-propagating wave amplitude B_j are assumed. The standing P_s and reflected P_r pressure waves can be expressed in Equations (7.49) and (7.50), respectively, as

$$P_s = \sum_{j=0}^{\infty} A_j e^{i\left(\omega t - K^4 x_j\right)} \tag{7.49}$$

$$P_r = \sum_{j=0}^{\infty} B_j e^{i\left(\omega t + K^4 x_j\right)} \tag{7.50}$$

where j is the junction number, $K = k/k_b$ is the wavenumber ratio, $k = \omega/c$ is the wavenumber, ω is the angular frequency, $c = 330$ m/s is the sound speed in air, and k_b is the bending wavenumber of the laminated composite pipe.

7.2.1.2.3 Acoustic properties of composite laminated muffler

The laminated muffler examined in this case study is made of basalt fiber-reinforced polymer composites. Due to the low strength and stiffness characteristics in the transverse direction of the laminate, it does not consist solely of the unidirectional lamina.

Consequently, some laminae in most laminates are placed at an angle. Therefore, the stress–strain relationship for an angled lamina must be developed. The new axes are referred to as local axes in the 1–2 coordinate system, where direction 1 is parallel to the fibers and direction 2 is perpendicular to the fibers. The angle between the local axes in the 1–2 coordinate system and the global axes in the x–y coordinate system is known as the fiber angle θ. The plane stress-transformed reduced stiffness coefficients $\overline{Q}_{ij}$ of the lamina can be expressed in terms of engineering notations (see Figure 7.56) as follows:

$$Q_{ij} = \begin{bmatrix} Q_{11} & Q_{12} & Q_{13} \\ Q_{12} & Q_{22} & Q_{23} \\ Q_{13} & Q_{23} & Q_{66} \end{bmatrix} = \begin{bmatrix} \dfrac{E_{11}}{(1-v_{12}v_{21})} & \dfrac{v_{21}E_{11}}{(1-v_{12}v_{21})} & 0 \\ \dfrac{v_{21}E_{11}}{(1-v_{12}v_{21})} & \dfrac{E_{22}}{(1-v_{21}v_{12})} & 0 \\ 0 & 0 & G_{12} \end{bmatrix} \tag{7.51}$$

$$\overline{Q}_{11} = Q_{11}\cos^4\theta + 2(Q_{12}+2Q_{66})\sin^2\theta\cos^2\theta + Q_{22}\sin^4\theta \tag{7.52}$$

$$\overline{Q}_{12} = (Q_{11}+Q_{22}-4Q_{66})\sin^2\theta\cos^2\theta + Q_{12}(\sin^4\theta+\cos^4\theta) \tag{7.53}$$

$$\overline{Q}_{22} = Q_{11}\sin^4\theta + 2(Q_{12}+2Q_{66})\sin^2\theta\cos^2\theta + Q_{22}\cos^4\theta \tag{7.54}$$

$$\overline{Q}_{66} = (Q_{11}+Q_{22}-2Q_{12}-2Q_{66})\sin^2\theta\cos^2\theta + Q_{66}(\sin^4\theta+\cos^4\theta) \tag{7.55}$$

The bending stiffness D_{ij} can be calculated from

$$D_{ij} = \frac{1}{3}\sum_{n=1}^{N}\left[(\overline{Q}_{ij})\right]_n (h_n^3 - h_{n-1}^3), \quad i,j = 1, 2, 3, \ldots \tag{7.56}$$

Using the above analysis of composite material, it can be concluded that the muffler's performance is greatly influenced by the stacking sequence of laminating plies and the fiber angle θ. When the muffler is excited by sound waves, a relationship exists between the sound pressure inside such a DCLCM and the normal vibration velocity [431].

7.2.1.2.4 Acoustic transmission loss
In mufflers, the theoretical definition of transmission loss is the logarithmic ratio of incident power to transmitted power in the case of reflection-free terminations. This can be expressed in terms of sound pressure by solving Equation (7.57):

$$P_j(x,t) = \sum_{j=0}^{\infty} A_j e^{i(\omega t - K^4 x_j)} + \sum_{j=0}^{\infty} B_j e^{i(\omega t + K^4 x_j)} \tag{7.57}$$

where k_b is the bending wavenumber of the laminated composite pipe:

$$k_b{}^4 = \frac{m\omega^2}{\sin^4\theta\left(D_{11}\cos^4\theta + 2(D_{12} + 2D_{66})\sin^2\theta\cos^2\theta + D_{22}\sin^4\theta\right)} \tag{7.58}$$

where $m = S_2/S_1$ is the area ratio.

Figure 7.64 presents the algorithmic flowchart of Equation (7.57), a solution technique to find the acoustic pressure P_j wave in each junction (j) of the DCLCM model (see Figure 7.53). By applying algorithms of continuity of pressure (P) in Equation (7.59) and volume velocity (U) in Equation (7.60) and substituting with boundary conditions at each junction (j) in the DCLCM model based on the flow condition inside the muffler (pt_1– pt_8) and section area (S_j) (see Figure 7.53):

$$P_{s_j} + P_{r_j} = P_{s_{(j+1)}} + P_{r_{(j+1)}}, \quad (\text{BC})_j \text{ based on } (pt_1 \sim pt_8), \ S_j \tag{7.59}$$

$$U_{s_j} + U_{r_j} = U_{s_{(j+1)}} + U_{r_{(j+1)}}, \quad (\text{BC})_j \text{ based on } (pt_1 \sim pt_8), \ S_j \tag{7.60}$$

where j is the junction number and U denotes the axial acoustic velocity, which can be obtained by the momentum equation:

$$i\rho\omega U = \frac{\partial P}{\partial x} \tag{7.61}$$

where ρ is the fluid density and $i = \sqrt{-1}$ is the imaginary unit. Then, the orthogonal properties of the eigenfunctions are used [432], where $L_1 = L_2 = L$ and $S_1 = S_3 = S_5$, $S_2 = S_4$, and the junction boundary condition is applied. Consequently, the resultant sound transmission loss of the DCLCM model is determined to be:

$$\text{TL} = 20 \log\left|1 + \frac{i\omega m \cos\theta}{2\rho c}\left(1 - \left(\frac{k}{k_b}\right)^4\right)\right| \tag{7.62}$$

Artificial neural networks are among the most widely used algorithms in artificial intelligence for solving advanced nonlinear problems [433, 434]. Each individual network within an artificial neural network consists of a number of computational nodes, with each node responsible for processing inputs and transferring the results of input calculations to output connections. The output of each node may serve as an input to one or more other nodes. Weights and biases are applied to scale and adjust the outputs, respectively, for example, in the function $y = mx + b$, y and x refer to output and new output, respectively, m is the weight, and b is the bias. Some networks are activated to determine the output with the type of function being linear or nonlinear by adding activation functions. The ReLU is the most commonly used activation function in deep learning [435].

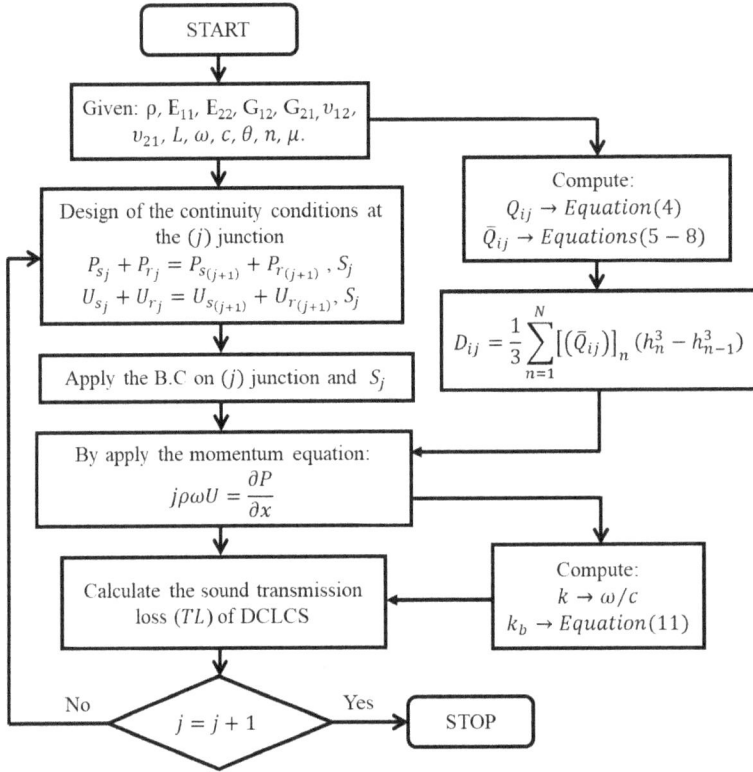

Figure 7.64: Algorithm of the governing equation solution technique to find the muffler's transmission loss.

7.2.1.3 Artificial neural networks

Before adjusting the weights and biases in an artificial neural network, the model does not work well, that is, the artificial neural network model is not trained. The learning of neural network can be achieved automatically from raw data as a hierarchical feature representation [426, 436] or can be trained via case study examples. In the models designed for this case study, we used supervised learning to train the network by comparing the training data and the model's predictions to the actual data. The training accuracy can be improved by updating the trainable parameters to optimize the error between prediction and actual data.

The type of artificial neural network can be determined based on the input data, as demonstrated in this case study. In this instance, artificial neural network models are used to analyze the time-series data.

In this case study, the two major types of artificial neural network models for analyzing time-series data are the RNN-LSTM and the CNN.

7.2.1.3.1 Convolutional neural network

CNNs are used to analyze groups of data, such as time series, images, sentences, sound recordings, and more. Weight matrices in CNNs are applied as kernels or filters to extract features [437, 438].

As shown in Figure 7.65, a proposed CNN model typically consists of feature extraction through a stack of layers on the input layer, such as convolution, activation, and pooling, and classification through fully connected layers for outputting the scores for each class. Each layer is responsible for different functions and uses the result from the previous layer as its input. Equation (7.2) describes the operation of the proposed CNN.

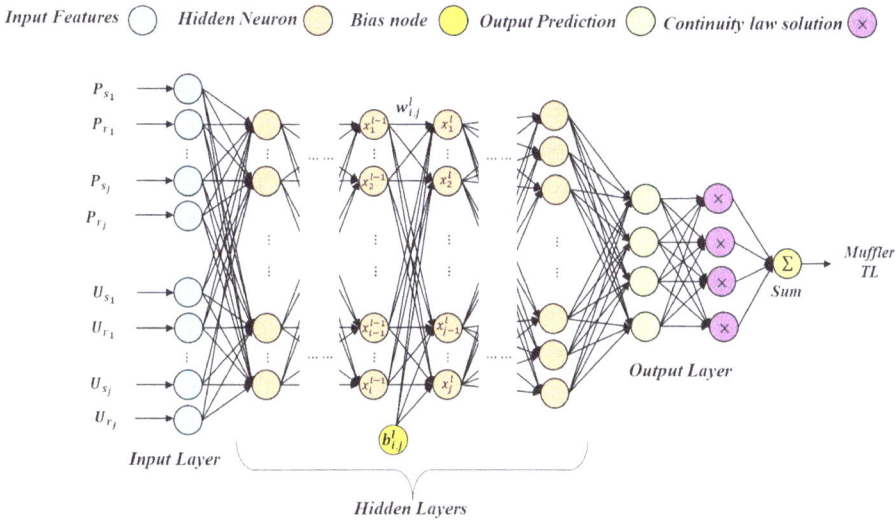

Figure 7.65: The architecture of the proposed convolutional neural network with a fully connected network.

7.2.1.3.2 Recurrent neural network with long short-term memory (RNN-LSTM) blocks

RNNs deeply analyze time-series data by applying feedback loops to the original artificial neural network [439]. The biggest disadvantage of RNNs is the vanishing gradient problem, where, during the backpropagation process, the error signal used to train the network exponentially decreases the further you travel backward in RNN; thus, sometimes computational nodes known as LSTM blocks are used to relieve this problem, as shown in Figure 7.66. Data feature extraction is performed from the first layers of the artificial neural network. These layers are responsible for extracting significant information from the input data [440].

LSTM blocks are a special type of RNN with a gating mechanism and memory cells, which greatly improve the performance of RNNs. There are three types of gates within each LSTM cell: the input gate, forget gate, and output gate. These gates define the state of each memory cell by using a sigmoid activation function, allowing information to be transmitted selectively. The memory cell, which retains the long-term state c_t is the key architecture of each LSTM cell. The internal architecture of a single LSTM cell is shown in Figure 7.67.

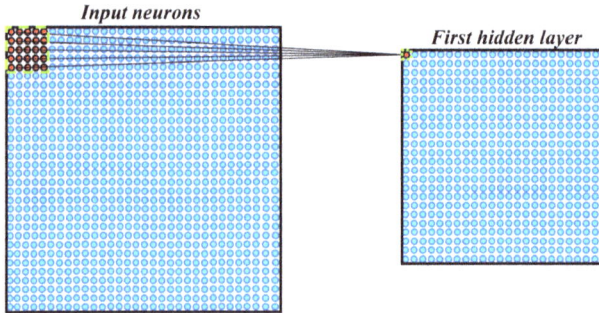

Figure 7.66: A 5 × 5 filter rolling around an input volume and generating an output.

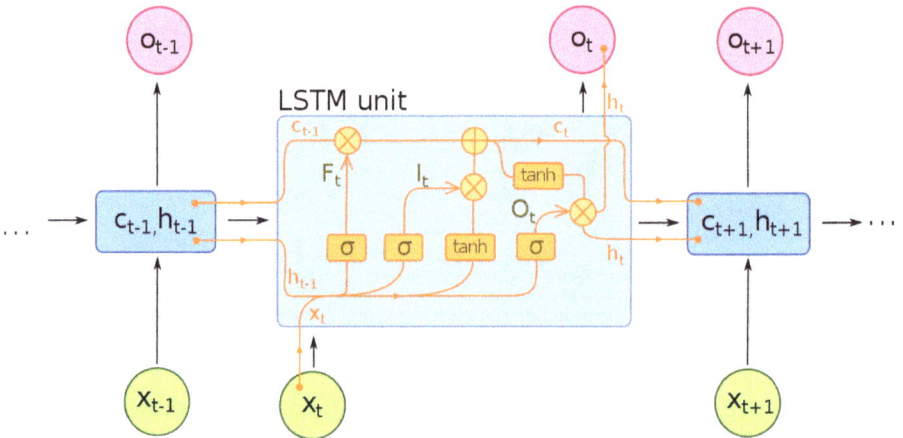

Figure 7.67: A single block diagram of an RNN-LSTM [41].

We can describe the operation of the three gates presented in Figure 7.67 using Equations (7.63)–(7.65). Equations (7.66)–(7.68) define the cell states c_t and hidden states h_t of each LSTM block unit at time t:

$$i_t = \sigma(W_i \cdot [h_{t-1}, x_t] + b_i) \tag{7.63}$$

$$f_t = \sigma(W_f \cdot [h_{t-1}, x_t] + b_f) \tag{7.64}$$

$$o_t = \sigma\left(W_o \cdot [h_{t-1}, x_t] + b_o\right) \tag{7.65}$$

$$c'_t = \tanh\left(W_c \cdot [h_{t-1}, x_t] + b_c\right) \tag{7.66}$$

$$c_t = f_t \odot c_{t-1} + i_t \odot c'_t \tag{7.67}$$

$$h_t = o_t \odot \tanh(c_t) \tag{7.68}$$

where W_f, W_i, W_c, and W_o represent the weight matrices of LSTM blocks; b_f, b_i, b_c, and b_o denote the bias vectors of the LSTM blocks; f_t, i_t, and o_t are the forget gate, input gate, and output gate vectors at time t; c_{t-1} and c_t mean, respectively, the previous cell state and a new candidate value. $\sigma(z)$ and $\tanh(z)$ are utilized as the activation functions, as follows:

$$\sigma(z) = \frac{1}{1 + e^{-z}} \tag{7.69}$$

$$\tanh(z) = \frac{e^z - e^{-z}}{e^z + e^{-z}} \tag{7.70}$$

7.2.1.3.3 Bayesian genetic algorithm optimization

Generally, there are several strategies for Bayesian genetic algorithm optimization modeling of objective functions (f), such as Gaussian processes [441–443], random forests [440], and tree-structured Parzen estimators [444, 445]. Figure 7.68 compares the grid search and Bayesian genetic algorithm optimization methods for tuning the model's hyperparameters. As shown in the figure, the yellow dot represents the assessment of the model in each method; note that the grid search method may explore the search space in a coarse manner, whereas Bayesian genetic algorithm optimization methods can test any possible combination within the space and intelligently suggest combinations to obtain optimal solutions with fewer evaluations. In this research, we applied the Bayesian genetic algorithm techniques for quick and intelligent tuning.

7.2.1.4 The work description

With the desire to explore alternative indirect monitoring acoustic behavior of DCLCM frameworks and inspired by deep learning, this case study discusses the framework of deep learning that analyzes transmission loss data of DCLCM to optimize its acoustic transmission loss. To evaluate the idea of the presented case study, the study is divided into the following sections:

Data collection: The ability and accuracy of the designed artificial neural network are based on the volume of data provided to the artificial neural network; this accuracy will be improved when there are more data for training. In the present case study, the data were computed using MATLAB from a DCLCM model. This dataset was used for training, validation, and testing for the artificial neural network.

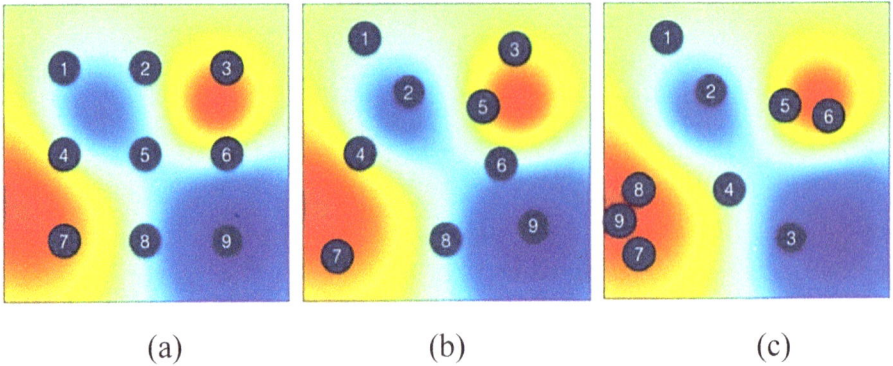

(a) (b) (c)

Figure 7.68: Comparison between (a) grid search, (b) random search, and (c) Bayesian genetic algorithm optimization techniques for tuning the model's hyperparameters.

The established algorithm: From the data collected in the previous step, via Python, an RNN-LSTM and a CNN are developed. Then, using a Bayesian genetic algorithm, we train the parameters and tune the models.

Figure 7.69 presents the flowchart of the framework for the proposed artificial neural networks used in this case study. As shown in Figure 7.69, the weight coefficients between the processing neurons are adjusted to achieve a desired goal during the training process. The neurons are interconnected via feed-forward links, which are considered for basic neuron computation. The multiplication results of the neuron inputs and the connection weights between the input layer and the hidden layer are obtained. Then, a bias value is added to these results using Equation (7.2).

The results of the calculation processes are subjected to an activation function, which generates the neuron output results. Similar operations can be applied to the output layer. The activation function determines the input/output behavior of the network. The error, known as the difference between the artificial neural network's output and the desired output, is calculated. Finally, new weight values are obtained. This method must be repeated until the error is acceptable. The training method is continued for all the data in the training dataset.

7.2.1.5 Validation of the proposed method

In this section, a convergence investigation is conducted for the proposed method. The acoustic transmission loss at all octave band center frequencies is calculated and compared with available experimental results in the literature. Table 7.20 presents a convergence and comparison study for a muffler structure optimization design aimed at eliminating noise in the pumping system, as described by Liu et al. [446]. The muffler is installed at the outlet of the pump pipeline system. The muffler material is commercial steel, and the structure of the muffler is shown in Figure 7.70.

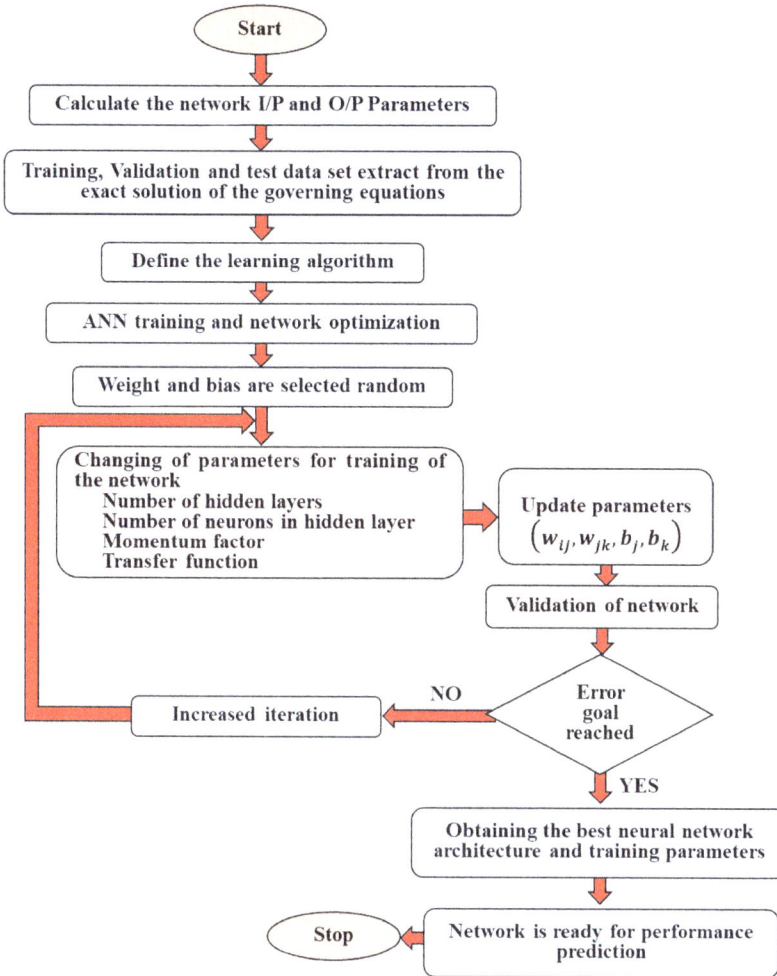

Figure 7.69: The flowchart of the artificial neural network model framework.

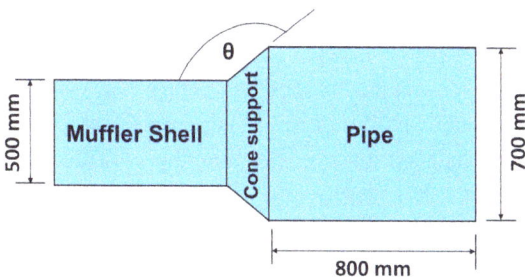

Figure 7.70: The muffler structure.

The experimental test device used by Liu consists of a booster pump, motor, valve, pressure transmitter, flowmeter, muffler, data acquisition system, and other components. They studied the noise reduction performance using three values for the expansion angle of the flow channel (θ): 120°, 145°, and 160°. They found that when the extension angle was 145°, the muffler had the best sound attenuation effect. In this validation, the measured data for the muffler with an expansion angle of 145° were compared with the simulation data from the proposed method. As shown in Table 7.20 and Figure 7.71, we can see that the acoustic transmission loss at all octave band center frequencies obtained from muffler tests followed the same trend as the calculations from the proposed method.

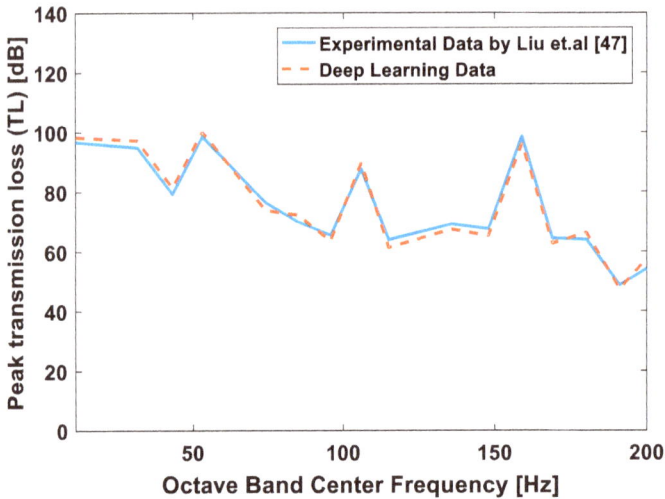

Figure 7.71: Comparison between the proposed method and experimental values of peak transmission loss for the muffler.

7.2.1.6 Data collection

In this case study, the dimensions of the DCLCM model and all S/C ratios (B/L) in most of the calculations in the analytical section are included in Table 7.19.

A MATLAB model for the DCLCM representation, as shown in Figure 7.62, was run at various S/C ratios to generate simulated examples of the DCLCM system response. The simulation solves the system's acoustic equations to determine the acoustic pressure at each junction of the DCLCM for every pressure and volume velocity at each junction in the DCLCM system. The simulation inputs are derived from Figures 7.72 and 7.73, and their accompanying derivations are presented in Equations (7.56)–(7.70).

The acoustic transmission loss properties of the proposed muffler are illustrated in Figures 7.72 and 7.73 (for B/L = 0, 1.0, 0.2, 0.4, 0.6, 0.8, 1.0), where high attenuation is

Table 7.20: The peak transmission loss of the muffler expansion angle of the flow channel at 145° along the frequency range.

Octave band center frequency (OBCF) (Hz)	Expansion angle of the flow channel at 145°	
	Liu et al. [47]	Proposed method
11.1	96.5	98.2
31.5	94.7	97.1
43.2	79.2	81.4
53.1	98.4	100
74.4	76.3	73.7
84.8	69.9	72.2
95.7	65.4	63.5
106	87.5	89.6
115	63.9	61.3
136	69.1	67.4
148	67.5	65.2
159	98.5	96.1
169	64.3	62.6
180	63.9	66.2
191	48.6	47.3
200	54.1	57.5

evident over a wide frequency range. Figure 7.72 shows the acoustic transmission loss distribution across all octave band center frequencies.

The DCLCM behavior is presented in Figure 7.72, where the acoustic transmission loss distribution changes over the frequency range with the change in muffler dimensions. Therefore, we found that the expansion chamber without a muffler (S/C ratio equal to zero) has equal domes, and the higher the S/C ratio, the more it changes the domes to an unequal shape, and the first dome has the smaller amplitude and frequency band than another one at each muffler geometrical configuration in S/C ratios presented in Table 7.19. The rate of rise and fall of the domes increases the acoustic transmission loss with the increase in frequency so that it reaches a peak at a certain frequency which then decreases and so on. In addition, the S/C ratio (B/L) in the muffler geometry has an effective effect in sound attenuation of transmission loss value, and we found that the increase in the S/C ratio (B/L) has an effect on widening the second dome and tend to cover two domes or more of the acoustic transmission loss for the lower S/C ratios. The cutoff frequency at which the behavior of acoustic transmission loss operates is 3,000 Hz. At all S/C ratios (B/L) presented in Table 7.19, the muffler's effect on the transmission loss value diminishes almost sequentially at frequencies of 130, 615, 825, 1,440, 1,780, 2,270, and 2,600 Hz, respectively.

Figure 7.73 shows the acoustic transmission loss distribution with respect to the S/C ratio over one complete period of frequency, ranging from 0 to 825 Hz. As shown in Figure 7.73, at a stationary frequency, the value of acoustic transmission loss gain changes with the increase in the *S/C* ratio (*B/L*), which indicates that the acoustic transmission loss has a high sensitivity to *S/C* ratio (*B/L*) changes. For the consequently transmission loss y, we can consider that the *S/C* ratio is the main parameter of frequency shifting, and that it is the powerful key in the acoustic transmission loss or gain.

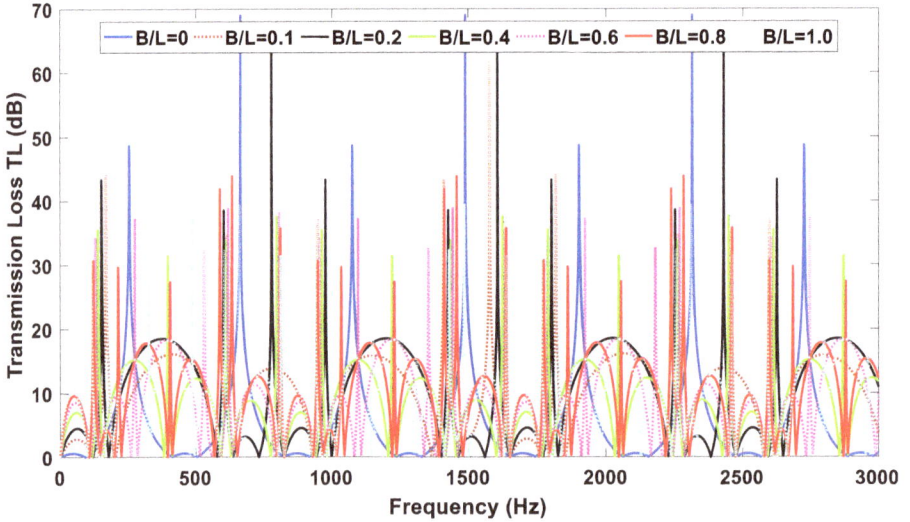

Figure 7.72: The transmission loss of DCLCM with different S/C ratios (B/L).

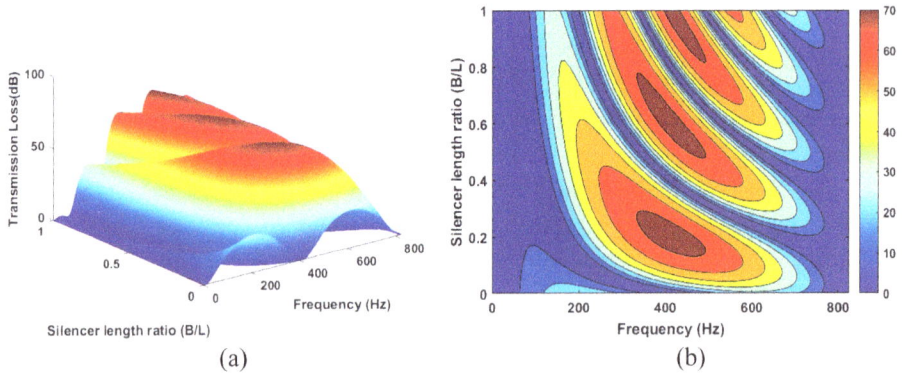

(a)   (b)

Figure 7.73: Modeled transmission loss versus *S/C* ratio (*B/L*) and full periodicity of frequency: (a) a 3D plot and (b) a contour plot.

7.2.1.7 The established algorithm

The establishment, training, and evaluation of a CNN or an RNN were conducted using Python software. Each neural network was developed via the same steps listed in Algorithm 7.3, based on steps discussed in Figure 7.69.

Algorithm 7.3: Training and evaluating of convolutional neural network and recurrent neural network.

1: **algorithm** Convolutional Neural Network or Recurrent Neural Network
2:   **input**: d: Transmission loss dataset, l: S/C ratio (B/L), W: Network parameter matrix weight w_{ij}, w_{jk} and bias b_j, b_k
3:   **output**: score of Convolutional Neural Network or Recurrent Neural Network trained model on test dataset to predict Transmission loss for various S/C ratio (B/L)
4:   **let** f be the feature set 3d matrix
5:   **for** i in dataset **do**
6:   **let** f_i be the feature set matrix of sample I
7:   **for** j in i **do**
8:     $v_j \leftarrow$ vectorize $_{(j, w)}$
9:     **append** v_j to f_i
10:    **append** f_i to f
11:   $f_{train}, f_{test}, l_{train}, l_{test} \leftarrow$ split feature set and prediction into train subset and test subset
12:   $M \leftarrow$ Convolutional Neural Network (f_{train}, l_{train}) or Recurrent Neural Network (f_{train}, l_{train})
13:   score $\leftarrow$ evaluate (I, l_{test}, M)

7.2.1.8 Development of artificial neural network models

In this case study, as shown in Figure 7.74, the deep learning neural network used mainly has an input layer of the muffler characteristics data, and three 1D convolutions' full connection (FC) layers for training each layer are 56×128, 28×256, 14×512, respectively. For two subsampling LSTM block layers, each layer is 14×512, 7×512, respectively. Then all the resultant 2D arrays from the pooled feature maps are converted into a single long continuous linear vector in a flattening layer that has 25,088 elements in one linear vector, and a softmax layer as the activation function in the output layer that predicts a multinomial probability distribution of muffler characteristics, such as transmission loss datasets.

In this case study, both RNN and CNN models have similar configurations, except for the region near the input layer used for feature extraction maps. The cost and accuracy of both models are evaluated, and the test set accuracy is evaluated both before and after training. The Bayesian genetic algorithm technique was used to optimize the training speed and accuracy.

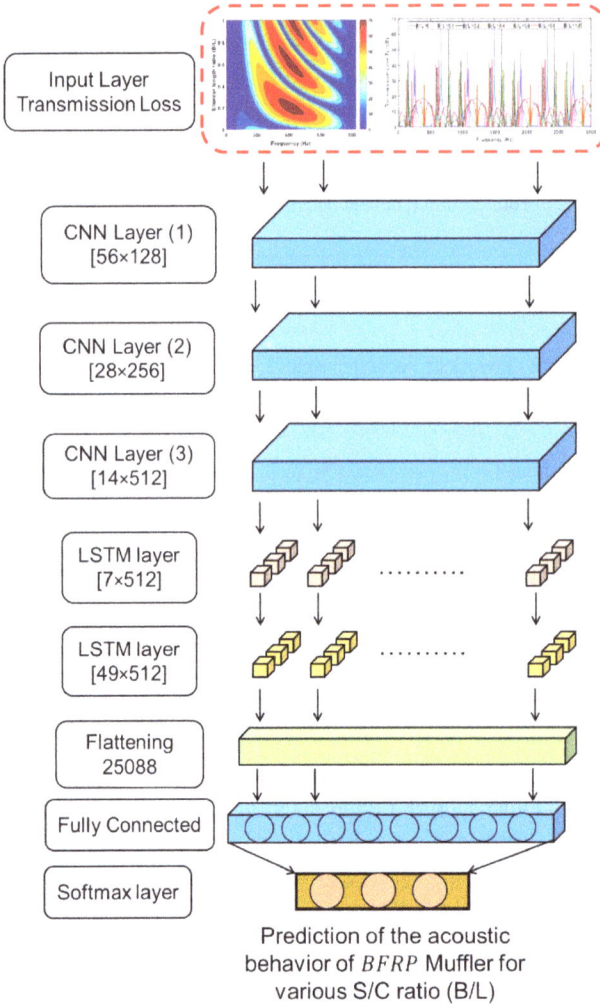

Figure 7.74: The three 1D convolutions' full connection (FC) layers, two subsampling long short-term memory block layers, and a softmax layer form a deep neural network.

7.2.1.9 Performance evaluation of artificial neural network models

Figure 7.75 shows the final performance of both models. As shown in Figure 7.75, which presents accuracy curves during the training of both the training and validation datasets, both CNN and RNN achieve accuracy above 90% on the test and validation datasets, while also saving sufficient time. Furthermore, we calculated the training and testing times for the RNN-LSTM and the CNN, as shown in Table 7.21.

In addition, the performance of CNN model is better than the RNN-LSTM model in terms of classification ability. The additional computations are what distinguish the performance of the CNN apart from RNN-LSTM. These computations are for feeding the hid-

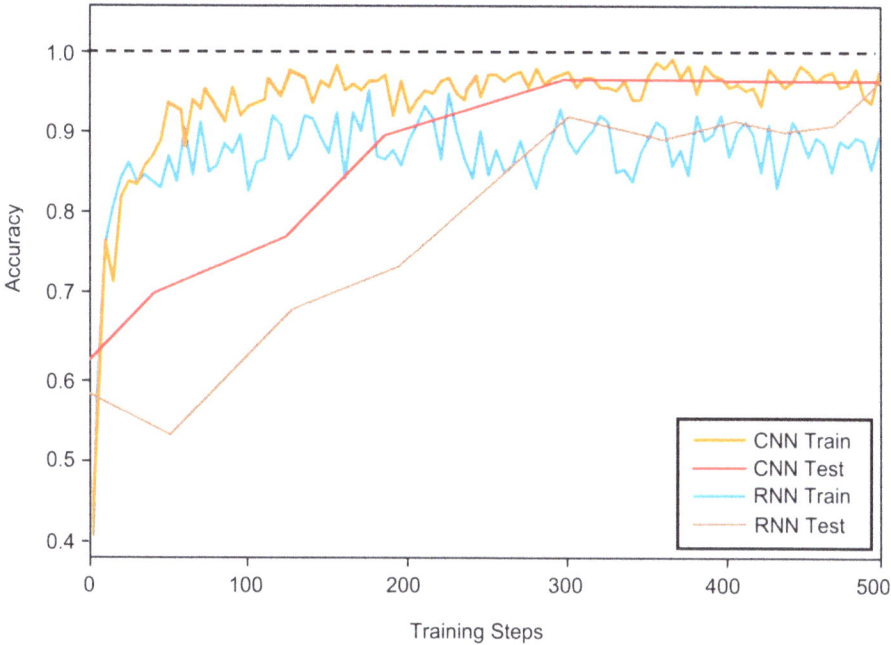

Figure 7.75: The sensitivity and accuracy of artificial neural network models during training.

den layer from the previous step to provide long-range contextual information to the next step. This means that the RNN-LSTM is incorrectly based on the built-up memory. On the other hand, the CNN is capable of training and classifying input data quickly and accurately because it extracts features within windows of time, including time-series data.

The drawbacks of CNN in the application of monitoring the acoustic behavior of DCLCM are the same as any time-series applications. The performance of a CNN is based on the dataset size and quality, but, in this case study, the dataset is small and free from noise. In addition, the poor random overfitting problem is one of the CNN's weaknesses, but, in this case study, this problem was not observed. As a result, for further understanding the weaknesses of CNNs, further research must be conducted using a more complex dataset. Simply, when directly compared with the fundamental equations of the convolution layer, we can observe a contrast in complexity. If the previous layer inputs have a large number of filters, the CNN model remains easier and faster to train and is simpler than the RNN-LSTM model.

7.2.1.10 Acoustic DCLCM geometry design optimization

For selecting the optimal DCLCM geometry design, a genetic optimization method is used. A genetic algorithm is a stochastic global search and optimization method that mimics the process of natural biological evolution. This process leads to the evolution

Table 7.21: Training times for RNN-LSTM and convolutional neural networks.

Model	Training time (s)	Testing time (s)
RNN-LSTM	87.45	0.1
Convolutional neural network	18.46	0.001

of populations of individuals that are better suited to their environment than the individuals from which they were derived, similar to natural adaptation.

Figure 7.76 illustrates the flowchart of a genetic optimization method applied in this case study. In this case, the DCLCM junction locations as a string of integer numbers are encoded analogous to the genetic code on a DNA string. Accordingly, the DCLCM acoustic outputs of transmission loss are propagated across different octave band center frequencies, which is analogous to the behavior of a breeding population for a number of individuals, each characterized by its DNA, and each individual is determined according to some fitness function. Here, it is preferable to raise individuals with high fitness (high octave band center frequencies) during the breeding process, so that useful genes are more likely to propagate through generations (high octave band center frequencies) and detrimental genes disappear (low octave band center frequencies). This is similar to the range of acoustic frequencies generated by DCLCM. Typically, a genetic algorithm code was written using MATLAB.

As the muffler development depends on optimizing its geometric design for high performance, and as transmission loss is an essential characteristic of the muffler, as well as the S/C ratio (B/L) having a high sensitivity to acoustic transmission loss, the (B/L) is therefore a powerful parameter in enhancing acoustic transmission loss. In fact, the (B/L) is the key parameter for muffler development.

In this section, we will use the simulated acoustic output generated from the deep learning analysis of transmission loss to maximize acoustic transmission loss. The optimal value of the dimension (B) is planned and carried out. Based on the final performance of both artificial neural network models used in this case study, the CNN model outperforms the RNN-LSTM model. Therefore, we will use the acoustic transmission loss output from the CNN model in the optimization process. It should be noted that the derivation processes of the maximum value of acoustic transmission loss were achieved using MATLAB software. Figures 7.77 and 7.78 show the results of these derivations, which were used to identify the optimal dimension (B) that maximizes the value of transmission loss.

Figure 7.78 shows the relationship between the values of transmission loss and the frequency range of DCLCM with the optimal value of S/C ratio (B/L), compared to a system without a muffler. As shown in Figure 7.78, the maximum value of transmission loss can be obtained at the DCLCM resonance frequency (612 Hz), which has been analytically estimated. Therefore, the maximum attenuation value is 60 dB. It is possible to conclude that the optimization method is the better way to obtain the maximum

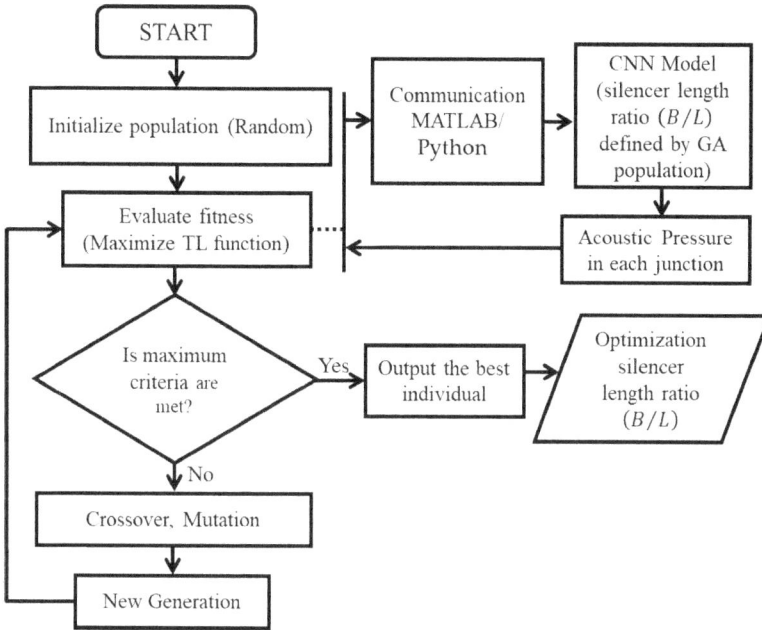

Figure 7.76: The flowchart of optimization methodology.

value of transmission loss corresponding to the defined dimensional range of Dual-Chamber Composite Silencer (DLCS). As a result of the DLCS dimension optimization, the optimal value of S/C ratio (B/L) is determined to be 0.106. This value corresponds to the highest level of sound attenuation in terms of transmission loss value.

7.2.1.11 Summary

This case study addresses the solutions to the drawbacks of experimental and numerical methods used to study the acoustic behavior of different muffler materials to improve their performance, where it is necessary to perform large, important, time-consuming calculations, particularly when the muffler is made from advanced materials such as composite materials. All of these problems were resolved in this study by using modern methods, specifically a deep learning algorithm, to predict the acoustic behavior of muffler materials, thereby saving effort and time in muffler design optimization. The acoustic behavior of a DCLCM made from basalt fiber-reinforced polymer laminated composite was predicted using Python through two types of deep neural network architectures: RNN-LSTM and CNN. A comparison between these two architectures was conducted in terms of the speed of training and accuracy of predictive data. The results showed that both the CNN and RNN-LSTM models achieved accuracy above 90% on test and validation datasets. Furthermore, the CNN model outperformed the RNN-LSTM model in terms of acoustic monitoring capability. The acoustic parameters, including transmission loss

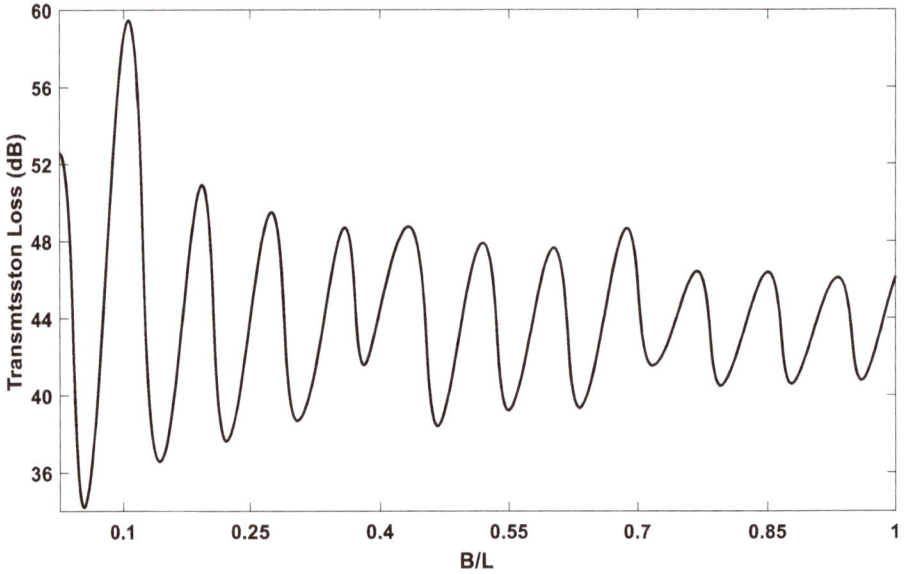

Figure 7.77: Transmission loss versus *S/C* ratio (*B/L*).

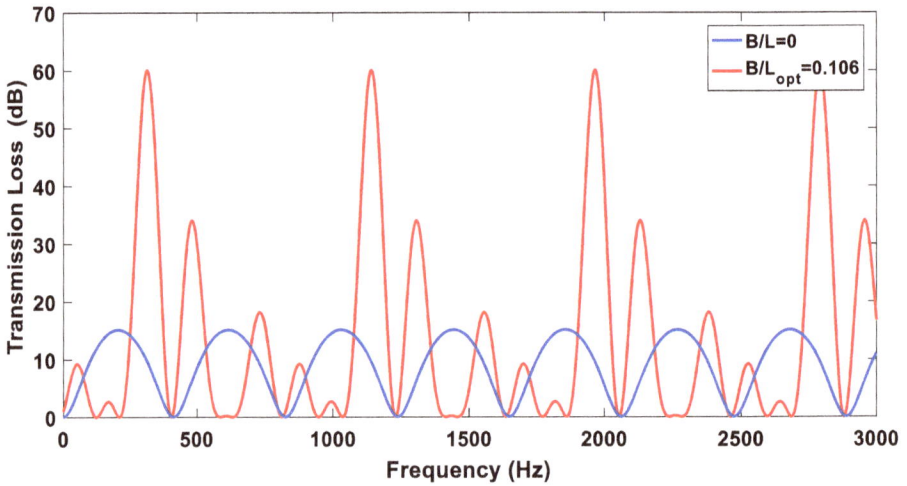

Figure 7.78: Transmission loss of DCLCM with the optimal value of the *S/C* ratio (*B/L*), compared with the system without a muffler $B/L = 0$.

and the power transmission coefficient, were computed by solving the exact solution of the governing acoustic equations of the muffler model in MATLAB. The model training parameters were optimized using Bayesian genetic algorithm optimization. The acoustic transmission loss output from the CNN model determined that the optimum muffler length is 0.0212 m, with a fixed pipe radius, achieving a maximum acoustic transmission loss of 60 dB at the resonance frequency of the DCLCM (612 Hz). This case study will reinforce the muffler industrials in the future, and its design may one day be equipped with deep learning-based algorithms.

7.3 Structural health monitoring of composite plates

Composite plates, with their high strength-to-weight ratio and customizable properties, have diverse applications across industries, including aerospace, automotive, construction, and marine, for structural components, decorative panels, and more. Structural health monitoring of composite plates involves the use of sensors and techniques to detect, locate, and quantify damage, ensuring the integrity of these structures, which are widely utilized in aerospace, civil, and marine engineering.

7.3.1 Case Study (5): structural health monitoring of composite plates using electrical capacitance sensors and system control theory

In this case study, a new framework for damage detection in composite plate systems made of basalt fiber-reinforced polymer is proposed through the analysis of electrical capacitance sensor measurements. First, the finite element model of the plate damage is established, where one side of the plate is subjected to the fatigue effects of applied pressure. Then, the electric potential difference between the electrode pairs is measured before and after the damage. The distributed electrical capacitance sensor measuring electrodes are installed around the plate and subjected to transient external excitations between each other. The transfer function of the "open-loop" plate system is proposed and applied to reflect damage evolution. The accuracy and reliability of the proposed technique are validated using available experimental results from the literature. Results show that the signal magnitude will suddenly and sensitively change when damage begins to grow under the electrode pairs, which shows the effectiveness of the proposed approach and its promising potential for engineering applications.

7.3.1.1 Methodology
The literature mentioned above examined the failure mechanisms, directives, design standards; and damage detection utilized the principles of forward, inverse electrical capacitance sensors, and introduced the detection sensitivity and the electronic design of electrical capacitance sensors. Also, the frequency function method was used for damage detection. However, to the best of the authors' knowledge, no prior literature has utilized the electrical capacitance sensor-based frequency function "system control theory" for composite plates, and this case study is the first study for that issue.

In this research, a novel diagnostic framework for composite plates is proposed by analyzing the exciting and measuring electrodes of electrical capacitance sensors as the inputs and outputs of a system. The damaged regions can be identified by applying transient external excitations between electrode pairs, while signal changes are observed in the electrodes located over the damaged regions. Subsequently, by deriving the transfer function of the "open-loop" composite plate system, the damage evolution is determined. The main motivation for this study is that, rather than using the structural responses such as vibration or acoustic signals, the electric potential difference is employed to reflect and analyze the variation in "system time and frequency characteristics" caused by different extents of structural damage. Figure 7.79 shows the proposed framework for damage detection in Composite plate.

7.3.1.2 The geometric model

The laminated composite square plate has a length of 280 mm with 15 mm height, and a stacking distribution of three plies is $[0/+45/-45]_S$. The thickness of each ply is 5 mm. The uniform pressure of plate is $P_0 = 15$ MPa, and the fatigue loading cycle number is $N_0 = 2.6E6$. The corresponding elastic modulus values were $E_1 = 93.5$ GPa, $E_2 = E_3 = 20$ GPa, and the shear modulus values were $G_1 = G_3 = 2.35$ GPa, $G_2 = 8.5$ GPa. The Poisson coefficients were v_1 $v_1 = v_3$ $v_3 = 0.28$, v_2 $v_2 = 0.3$, and the density was 2,700 kg/m^3. The electrical capacitance sensors consist of 12 rectangular electrodes fixed on both sides of the specimen surface, as shown in Figure 7.80. The electrodes are separated from each other by a 30 mm gap. The capacitance values are measured between each pair of electrodes using electrical capacitance sensors, and the equivalent node potential derived from the measured capacitance values is converted.

The numbering order of the electrode system is illustrated in Figure 7.81. Suitable finite elements were selected; that is, for simulating structural characteristics, SHELL99 element is used, whereas for simulating electrical characteristics, SOLID123 element is used. The main principle for detecting damage evolution is that variations in the dielectric constant of the plate material, corresponding to the damage evolution, cause changes in the capacitance measurements between the electrical capacitance sensor electrodes related to damage evolution under electrical capacitance sensor electrodes.

The S–N curve of the relationship between fatigue life N_f and fatigue stress (S_f) can be plotted in Figure 7.82. Using the power formula $S_f = aN_f{}^b$, we found it suitable to fit the data of the S–N curve by providing appropriate values for CF, which is equal to 0.9845, and a, b are the fatigue constants, equal to 1.69 MPa and -0.08944, respectively.

Figures 7.83 and 7.84 represent the geometrical model of the composite plate structure. The four ends of the composite plate are fully clamped supports (CCCC) (shown in blue), and electrical capacitance sensors are installed and distributed on both sides of the composite plate surface. In mathematics and signal processing, the Hilbert transform, a linear operator, is a basic tool in Fourier analysis and can be used to find the analytic signal of a given signal.

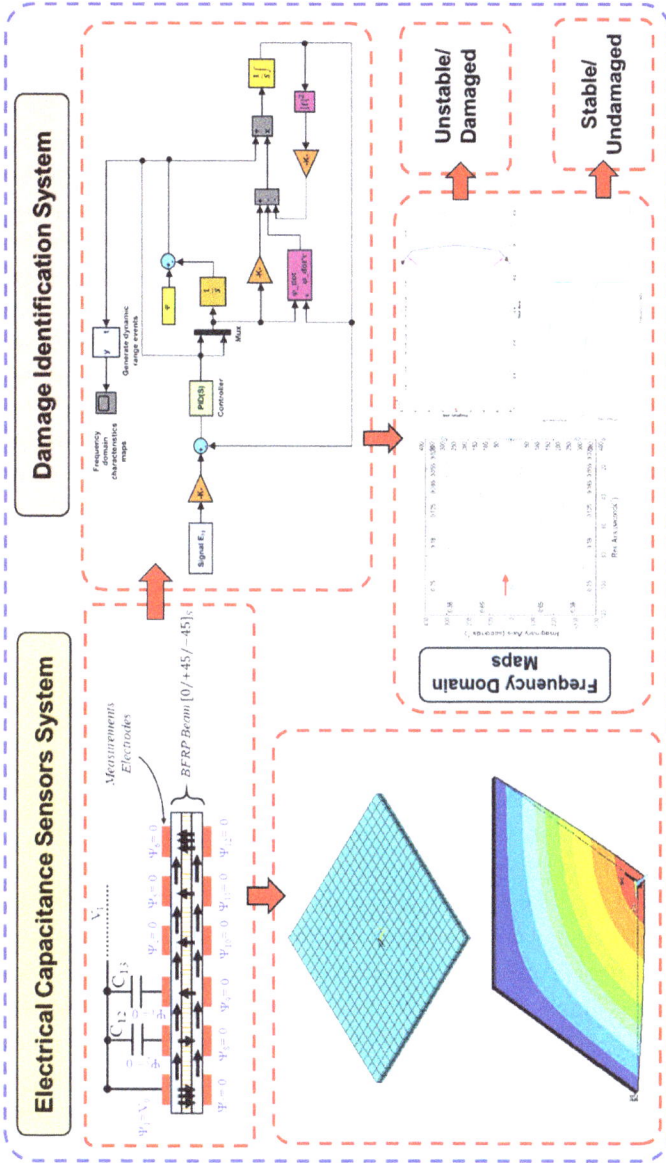

Figure 7.79: Illustration of the proposed damage detection.

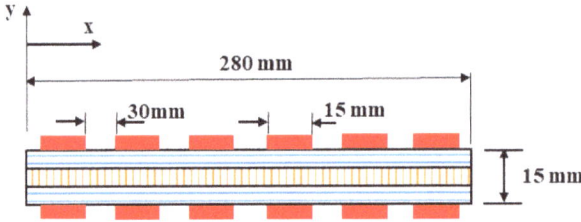

Figure 7.80: Geometric model diagram.

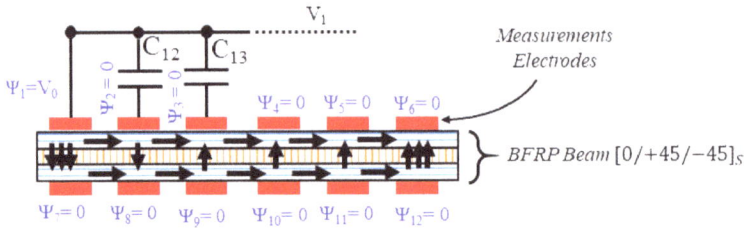

Figure 7.81: Schematic representation of the measurement principle of an electrical capacitance sensor.

Figure 7.85 shows the maximum displacement of the plate, where the maximum von Mises stress is represented in Figure 7.86. As shown in Figures 7.85 and 7.86, the most significant damage occurs near the clamped supports.

7.3.1.3 Damage monitoring of basalt fiber-reinforced polymer composite plate using electrical capacitance sensors

As shown in Figure 7.81, the main principle for damage monitoring is that variations in the dielectric constant of the plate material, corresponding to the damage, cause changes in the capacitance measurements between the electrical capacitance sensor electrodes related to the evolution of damage under the electrical capacitance sensor electrodes. In this case, we established a sinusoidal signal as the external excitations and modeled the capacitance distribution transient response, and the excitation voltage signal is $V = 60\sin(100\pi t)$ V. The measuring electrodes are kept at 0 V. The meaning electrode is the source from which the voltage signal is acquired. As shown in Figure 7.87, the amplitude difference in the voltage for electrode 2 is clear before and after damage. It can be observed that the signal magnitudes decrease after the damage starts, while the signals approach zero when the damage region is outside the electrode 2 region. Moreover, we labeled the damage states as $D0$ for no damage and $D1$ for damage. The flowchart of the damage monitoring process steps is shown in Figure 7.88.

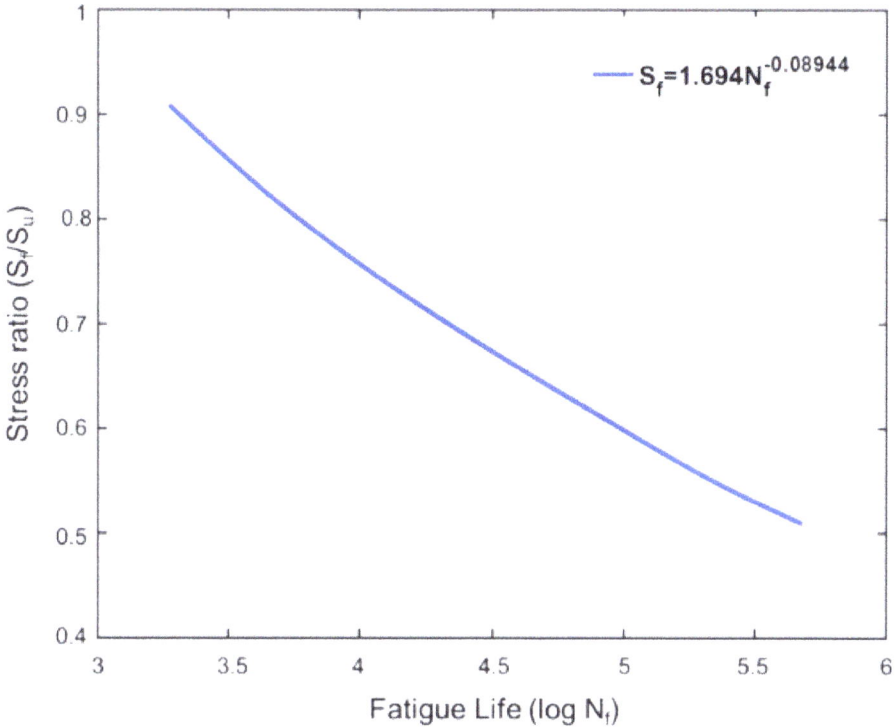

Figure 7.82: S–N curve of a basalt fiber-reinforced polymer composite plate.

7.3.1.3.1 Voltage signal analysis in the frequency domain

Considering the uncertainties in modeling, the typical fitting curves (average values) of the voltage variation response are selected (Equation (7.2)) to describe the output of the coupled field system excited by a sinusoidal input, where the cubic, quadratic, linear, and constant terms of the fitting equation are 0.0168, −0.0005, 0.0000, and 0.0000, respectively:

$$X_{o1}(t) = 0.01295t^3 - 0.0003t^2 \tag{7.71}$$

The single-input-single-output differential equation can be represented as follows:

$$a_0 x_o^{(n)}(t) + a_1 x_o^{(n-1)}(t) + \cdots + a_{n-1} x_o^{(1)}(t) + a_n x_o(t)$$
$$= b_0 x_i^{(m)}(t) + b_1 x_i^{(m-1)}(t) + \cdots + b_{m-1} x_i^{(1)}(t) + b_m x_i(t), n \geq m \tag{7.72}$$

where x_o is the system output and x_i is the system input

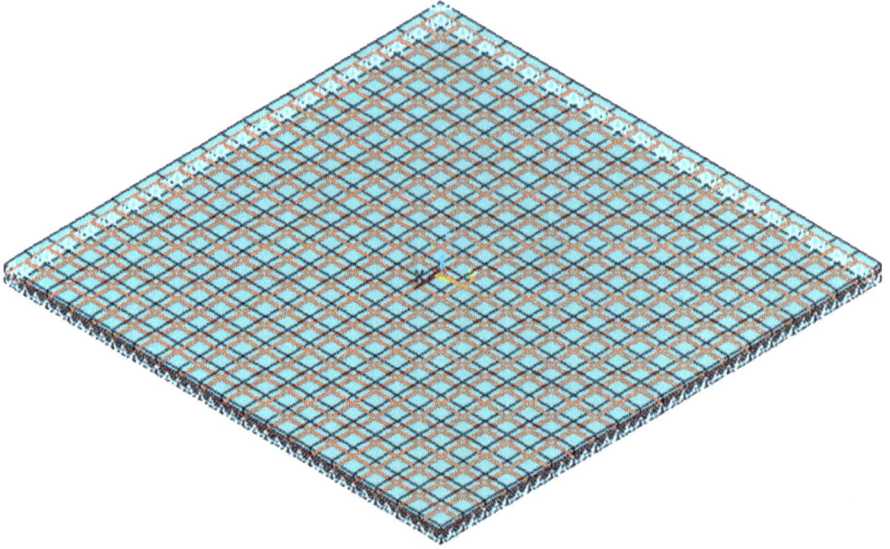

Figure 7.83: Structural–electrostatic coupled field modeling.

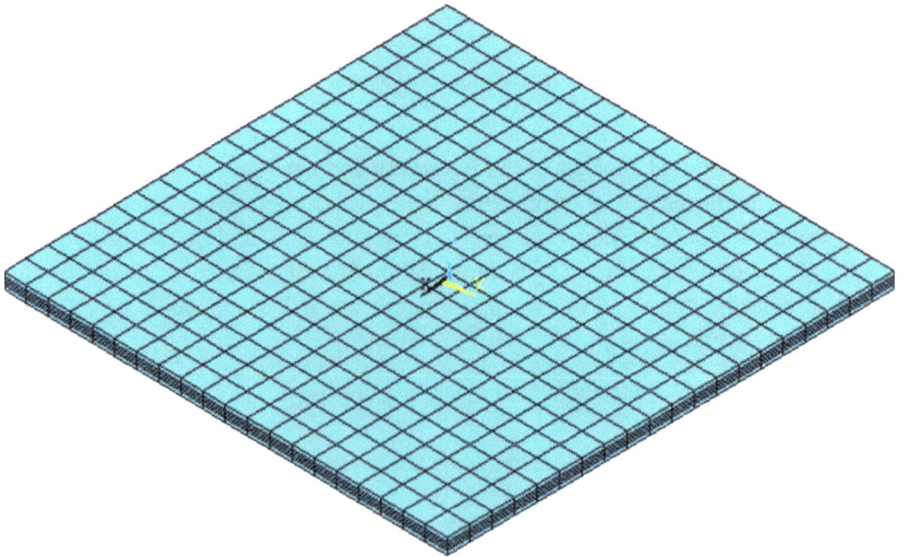

Figure 7.84: Finite element model.

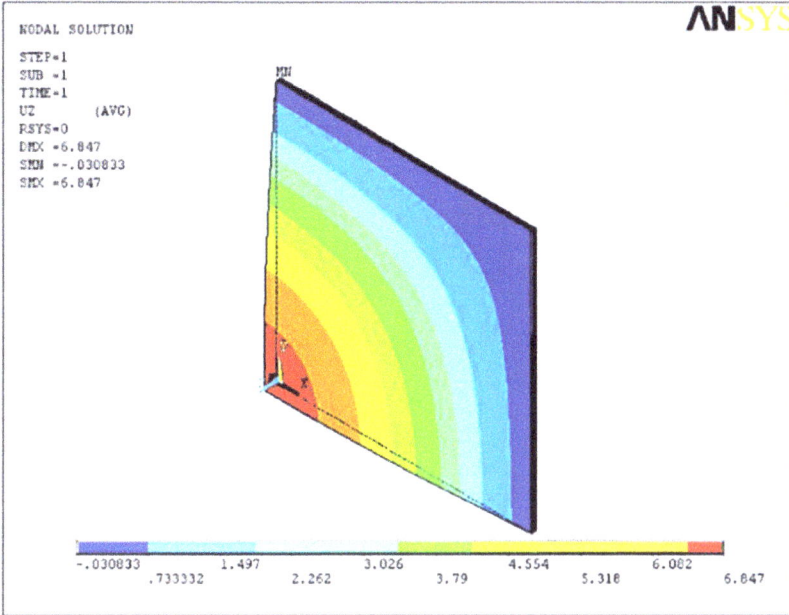

Figure 7.85: *Z*-component of displacement.

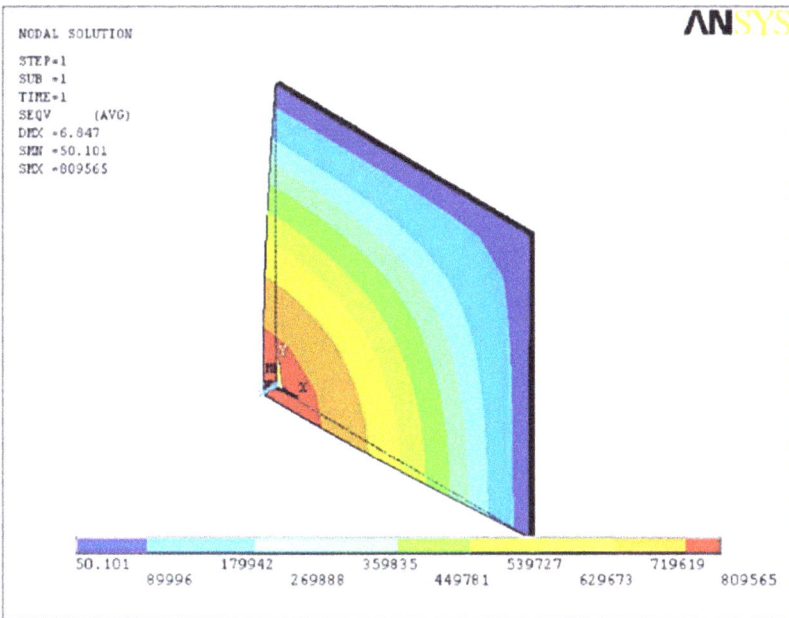

Figure 7.86: *Z*-von Mises stress.

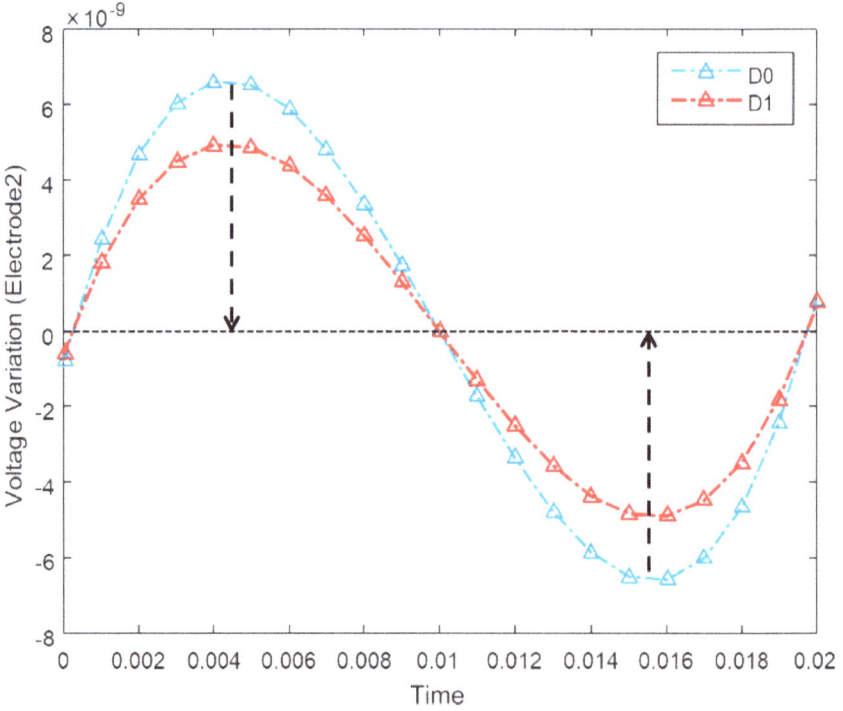

Figure 7.87: Transient response before damage $D0$ and after damage $D1$ accord under electrode 2.

- Dielectric: BFRP Composite plate
- Fatigue Loading
- Damage Index Change
- Modification of Stiffness
- Electric Field Distribution Change

Figure 7.88: Schematic diagram of damage monitoring.

$$G(s) = \frac{X_o(s)}{X_i(s)} = \frac{b_0 s^m + b_1 s^{m-1} + \cdots + b_{m-1} s + b_m}{a_0 s^n + a_1 s^{n-1} + \cdots + a_{n-1} s + a_n} \tag{7.73}$$

The system's input can be expressed as

$$X_i(t) = 60\sin(100\pi t), \quad X_i(s) = \frac{6{,}000\pi}{s^2 + 10{,}000\pi^2} \tag{7.74}$$

The system's output can be expressed as

$$X_{o1}(s) = \frac{30}{81 s^4} - \frac{1}{1{,}000 s^3} \tag{7.75}$$

The system's transfer function can be expressed as follows, based on the presented damage case:

$$G_1(s) = \frac{(s^2 + 10{,}000\pi^2)(81s - 30{,}000)}{2.5E9\pi s^4} \tag{7.76}$$

7.3.1.3.2 Frequency domain analysis methods

The plate system, subjected to different conditions of damage by fatigue has different response features when excited by a sinusoidal signal input. Let us consider it as a "control system" with an input excitation electrode and an output measurement electrode. When considering the system as an open-loop control system, this means that no feedback from output to the system, the main control on the system is from input only. The frequency domain analysis methods (e.g., zero-pole, Nyquist, Bode, and Nichols) are applied to measure the dynamic response characteristics of the system in relation to damage using Simulink software.

The zero-pole point map analysis for the presented damage case is shown in Figure 7.89. For the presented damage, we observe that the zero point of the system transfer function (100.8) moves to zero when the damage range increases. The Nyquist map analysis for the presented damage case is plotted in Figure 7.90. As shown in the figure, for the presented damage, we observe the variation in the endpoint trajectory of the vector $G(j\omega) = A(\omega)e^{j\varphi(\omega)}$ (which shows $A(\omega_i)$ appears the vector magnitude $G(j\omega_i)$ when the frequency equals ω_i, and the case of polar coordinates is $\varphi(\omega)$) when frequency ω changes from $0 \to \infty$. The Nyquist diagram shows that, in the nonoverlapped part, the system frequency of G is 2.02. The system frequency decreases as the damage range increases. The Bode map analysis for the presented damage case is plotted in Figure 7.91. As shown in the figure, for the presented damage, we can see the frequency of the excitation signal is 100π. Specifically, when damage occurs, the absolute value of the magnitude increases to over 200 dB, and the corresponding phase changes more than 180°. The Nichols map for the presented damage case is plotted in Figure 7.92. As shown in the figure, for the presented damage, the phase reaches about −69.5, accordingly G changes at the inflection point, with a gain value of − 475 at a frequency of 100π.

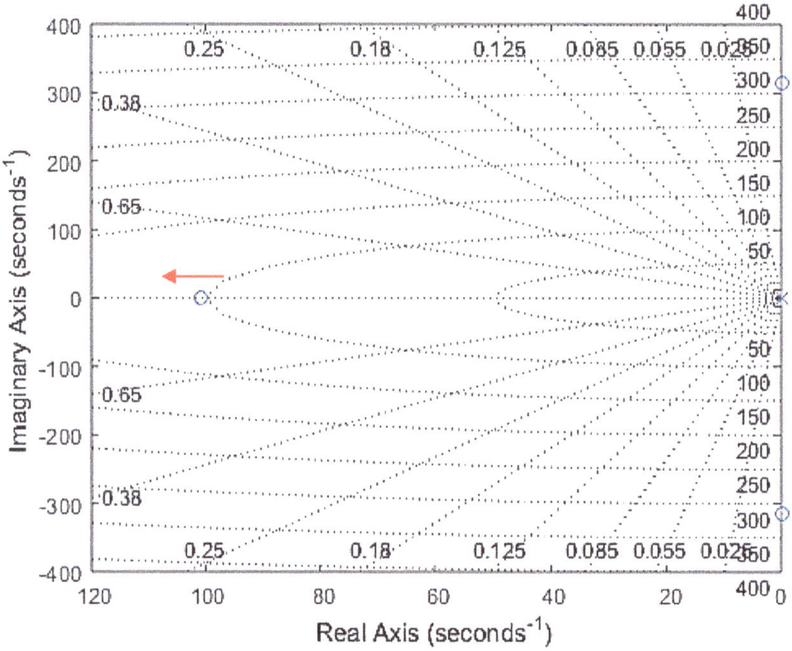

Figure 7.89: Pole-zero map for a basalt fiber-reinforced polymer composite plate.

The present theory depends on establishing the transfer function using input and output signals based on control system theory. The transfer function can evaluate the damage range in a specific area of the structure if the electrical capacitance sensor electrodes are excited and designed well, and this is done if the electrical capacitance sensor electrodes with a sinusoidal signal to analyze the system's dynamic performance (health status) through three different maps plotting frequency representations of system transfer function: Nyquist map, the Bode map, and the Nichols map. Each map reveals the relationship between the system's magnitude, phase, and frequency in distinct ways. The proposed approach is characterized by integrating control theory with damage mechanism theory in structural analysis.

7.3.1.4 Summary
In this case study, a finite element model of the basalt fiber-reinforced polymer plate damage system from the fatigue effect is established, by extracting the measurements of electrical capacitance sensor electrodes and loaded with a sinusoidal signal. Results show that the signal magnitude will suddenly and sensitively change when the dam-

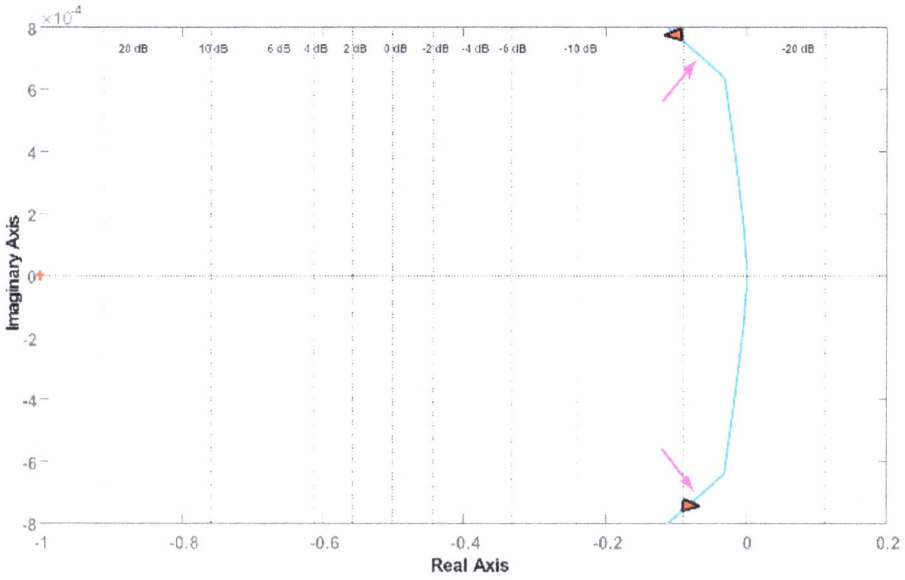

Figure 7.90: Nyquist map for a basalt fiber-reinforced polymer composite plate.

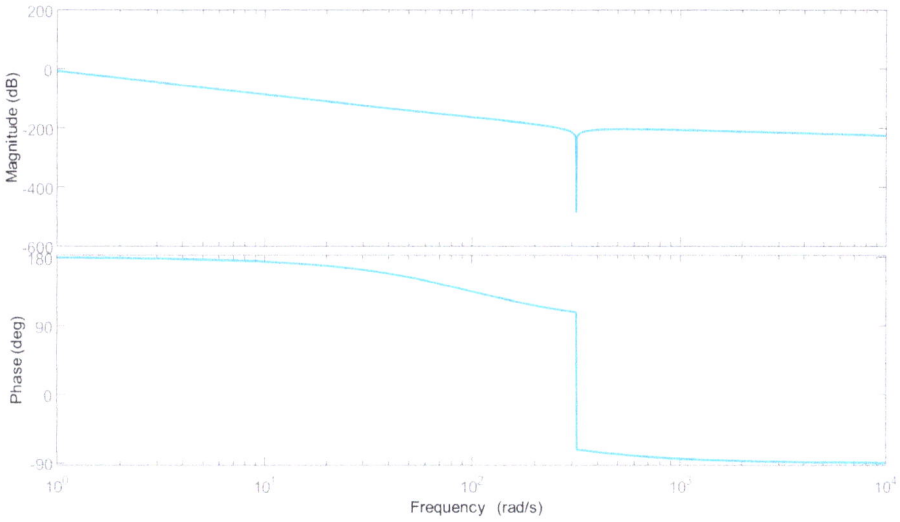

Figure 7.91: Bode plot for a basalt fiber-reinforced polymer composite plate.

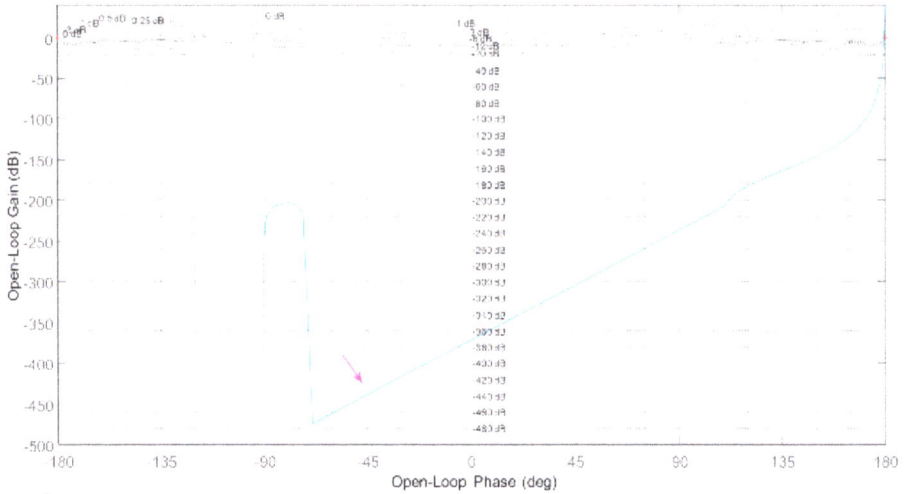

Figure 7.92: Nichols map for a basalt fiber-reinforced polymer composite plate.

age begins to grow under the electrode pairs. The plate system's transfer function and frequency domain are established and applied to reflect damage evolution by plotting the zero-pole point map, Nyquist map, Bode map, and Nichols map. The most significant conclusion of this case study is that, instead of using vibration signals from the structural system's responses, the variation in the electric field is used to observe the changes in time and frequency characteristics of the system caused by structural damage.

Notations

SHM	Structural health monitoring
FLD	Fatigue life diagram
UD	Unidirectional
RD	Relative damage
FRER	First to residual energy ratio
LMR	Leveraging the multipath reflected
CFDAC	Complex frequency domain assurance criterion
ODSs	Operational deformation shapes
CorV	Correlation function amplitude vector
HOMS	Higher-order mode shapes
PZ	Piezoelectric
PZT/PZTs	Piezoelectric transducer/transducers
PZS/PZSs	Piezoelectric sensor/sensors
NDT	Nondestructive testing
FBG	Fiber Bragg grating
AE	Acoustic emission
AI	Artificial intelligence
ANN/ANNs	Artificial neural network/networks
FFNN	Feed-forward neural network
RBNN	Radial basis neural network
BPNN	Backpropagation neural network
GRNN	Generalized regression neural network
LNN	Linear neural network
EBPTA	Error backpropagation training algorithm
IPV	Intelligent parameter varying
TMCMC	Transitional Markov chain Monte Carlo
GD	Gradient descent
DSD	Dynamic learning rate steepest descent
LFCR	Local frequency response ratio
EAs	Evolutionary algorithms
FRP	Fiber-reinforced polyester
CFRE	Carbon fiber-reinforced epoxy
CFRP	Carbon fiber-reinforced polyester
BFRP	Basalt fiber-reinforced polyester
GFRP	Glass fiber-reinforced polyester
WFC	Woven fiber composite
FEM	Finite element methods
PVDF	Polyvinylidene fluoride
PZWA	Piezoelectric wafer active
PZIA	Piezoelectric impedance transducers
PZFG	Piezo-floating-gate
SMART	Stanford Multiactuator–Receiver Transduction
CLT	Classical lamination theory
TRM	Time reversal method
MTRM	Modified time reversal method
SAMS	Single actuator and multiple sensors
VTRA	Virtual time reversal algorithm

https://doi.org/10.1515/9783112213094-008

ATROE	Alleviating time reversal operator effect
OTRP	Optimize time reversal excitation parameters
DBSCAN	Density-based spatial clustering applications with noise
WPDD	Weibull probability density distribution
FBG	Fiber Bragg grating
UDP	Undamaged pipe
DP	Damaged pipe
TFS	Time–frequency spectrogram
k-NN	k-Nearest neighbor
CNN	Convolutional neural network
TCNN	Traditional convolutional neural network
ECNN	Enhanced convolutional neural network
TPR	True positive rate
TNR	True negative rate
FPR	False positive rate
FNR	False negative rate
N	The number of electrodes
M	The number of independent capacitance measurements
C_{ij}	The interelectrode capacitance
Q_{ij}	The charge induced on electrode j when electrode i is excited with a known potential
V_{ij}	The potential difference between electrodes i and j
$\varphi(x,y,z)$	The potential distribution inside the ECS
$S(x,y,z)$	The sensitivity matrix
$E(x,y)$	The electric field vector
$p(x,y,z)$	The volume
$D(x,y)$	The electric flux density
$\varepsilon(x,y,z)$	The permittivity distribution
$\nabla.\varepsilon(x,y)$	The divergence of permittivity distribution
$\nabla\varphi(x,y)$	The gradient of potential distribution
S_j	A surface enclosing electrode j
ds	An infinitesimal area on electrode j
$\hat{n}$	The unit vector normal to S_j
ΔP	The pressure difference
P_1	The inlet or high-pressure point
P_2	The outlet or low-pressure point
h	The pipe thickness
D	The diffusion coefficient for the composite materials
k	The initial gradient of the water absorption curve
$M\%$	Mass of liquid absorption at a time (t)
M_∞	Mass of liquid absorbed at saturation
FDM	Fickian diffusion model
ECS	Electrical capacitance sensor
DNN	Deep neural network
EPD	Electrical potential difference
BCs	The boundary conditions
ε	The relative dielectric constant
S	The facing area
d	The distance between the capacitor plates
C	The potential difference between the capacitor plates

F_x, F_y	The local strength components in directions (x) and (y), respectively
F_{xy}	The local shear strength component
E_x, E_y, E_z	The elastic modulus in the "x," "y," and "z" directions
G_{xy}, G_{xz}, G_{yz}	The shear modulus in the "xy," "xz," and "yz" planes
v_{xy}, v_{xz}, v_{yz}	The Poisson's ratio in the "xy," "xz," and "yz" planes
K_x, K_y, K_z	The thermal conductivity in the "x," "y," and "z" directions, respectively
a_x, a_y, a_z	The thermal expansion coefficient in the "x," "y," and "z" directions, respectively
K_{xx}, K_{yy}, K_{zz}	The thermal conductivity in the "xx," "yy," and "zz" plane, respectively
$\varepsilon_x, \varepsilon_y, \varepsilon_z$	The electrical permittivity in the "x," "y," and "z" directions, respectively
R_x, R_y, R_z	The electrical resistivity in the "x," "y," and "z" directions, respectively
P_0	The applied pressure
P_{max}	The burst pressure
P_r	The ratio between the applied and the burst pressure
S_f	The fatigue stress
σ_{max}	The maximum fatigue stress
σ_{min}	The minimum fatigue stress
K_{max}	The maximum stress intensity
N_f	The fatigue life
E_m	The Young's modulus of matrix
V_m	The volume fraction of matrix
E_f	The Young's modulus of fiber
V_f	The volume fraction of fiber
E_c	The Young's modulus of composite
T_m	The polymer melting point
f^*	The fiber–matrix interface strength
S_{ult}	The ultimate tensile stress
R	The stress ratio
n	The percentage of drop in stiffness
x_i^l	The ith output map in layer l
x_i^{l-1}	The ith output map in layer l–1
w_{ij}^l	The weight matrix
b_j^l	The bias matrix
$f(\cdot)$	The nonlinear function that is applied component-wise
$R_f\%$	The long-term creep thermomechanical fatigue levels
$S_f(t)$	The long-term creep thermomechanical fatigue compliance
t	The creep time
CF	Correlation factor
FCGR	Fatigue crack growth rate
MSE	The mean squared error

References

[1]   Altabey, W.A., (2017). An exact solution for mechanical behavior of BFRP nano-thin films embedded in NEMS. Advances in Nano Research, 5(4), 337–357, https://doi.org/10.12989/anr.2017.5.4.337.

[2]   Altabey, W.A. (2017). A study on Thermo-Mechanical Behavior of MCD through Bulge Test Analysis. Advances in Computational Design, 2(2), 107–119, https://doi.org/10.12989/acd.2017.2.2.107.

[3]   Farrar, C.R. and Worden, K., An introduction to structural health monitoring. Phil. Trans. Royal Society Publishing A, 365, 303–315, 2007.

[4]   Diamanti K. and Soutis C., Structural health monitoring techniques for aircraft composite structures, Progress in Aerospace Sciences, 46, 342–352, 2010.

[5]   Ghiasi R., Ghasemi M.R., Noori M., Altabey W. A. (2019). A non-parametric approach toward structural health monitoring for processing big data collected from the sensor network, Structural Health Monitoring 2019: Enabling Intelligent Life-Cycle Health Management for Industry Internet of Things (IIOT) – Proceedings of the 12th International Workshop on Structural Health Monitoring (IWSHM 2019), September 10–12, 2019, Stanford, California, USA. http://dx.doi.org/10.12783/shm2019/32395.

[6]   Santoni G., Fundamental Studies in the LAMB-WAVE Interaction BETWEEN Piezoelectric Wafer Active Sensor and Host Structure During Structural Health Monitoring, PHD thesis, College of Engineering and Information Technology, University of South Carolina, 2010.

[7]   Vizzini A.G., Damage Detection in Blade-Stiffened Anisotropic Composite Panels Using Lamb Wave Mode Conversions, Master thesis, Arizona State University, 2012.

[8]   Su Z., Ye L., and Lu Y., Guided Lamb waves for identification of damage in composite structures: A review, Journal of Sound and Vibration, 295, 753–780, 2006.

[9]   Al-Tabey W.A., The Fatigue Behavior of Woven-roving Glass Fiber Reinforced Epoxy under Combined Bending Moments and Internal hydrostatic Pressure, PhD Thesis, Alexandria University, Egypt, 2015.

[10] Appropedia, Composites in the Aircraft Industry, http://www.appropedia.org/Composites_in_the_Aircraft_Industry (Accesses 1 Sep. 2022)

[11] Bannantine J.A., Comer J.J. and Handrock J.L., Fundamentals of Metal Fatigue Analysis. Prentice Hall, 1990.

[12] Talreja R., Damage and fatigue in composites – A personal account, Composites Science and Technology, 68, 2585–2591, 2008.

[13] Talreja R., Damage and fatigue in composites – A personal account, Composites Science and Technology, 68, 2585–2591, 2008.

[14] Schulte K., Baron C., and Neubert N., Damage development in carbon fibre epoxy laminates: cyclic loading, Advanced Materials Research and Developments for Transport, 281–288, 1985.

[15] Reifsnider K.L., Schulte K., and Duke J.C., Long-term fatigue behavior of composite materials, long-term behavior of composites, ASTM STP 813, American Society for Testing and Materials, 136–139, 1983.

[16] Nairn J.A., and Hu S., The initiation and growth of delaminations induced by matrix microcracks in laminated composites, Int. J. Fracture, 57, 1–24, 1992.

[17] Quaresimin M., and Susmel L., Multiaxial fatigue behaviour of composite laminates, J. Key Engineering Materials, 221–222, 71–80, 2002.

[18] Adden S., and Horst P., Stiffness degradation under fatigue in multiaxially loaded non-crimped-fabrics, Int. J. Fatigue, 32 (1), 108–122, 2010.

[19] Gude M., Hufenbach W., Koch I., and Protz R., Fatigue failure criteria and degradation rules for composites under multiaxial loading, Mechanics Composite Materials, 42(5), 443–450, 2006.

[20] Guedes R.M., Creep and Fatigue in Polymer Matrix Composites, Wood Head Publishing Series in Composites Science and Engineering, Woodhead Publishing, 366–405, 2011.

https://doi.org/10.1515/9783112213094-009

[21] Wu F., and Yao W., A fatigue damage model of composite materials, Int. J. Fatigue, 32, 134–138, 2010.

[22] Ferreira J., Reis P., Costa J., and Richardson M., Fatigue behaviour of Kevlar composites with nanoclay-filled epoxy resin, Composite Materials, 47(15), 1885–1895, 2012.

[23] Schulte K., Reese E., and Chou T.W., Fatigue Behaviour and Damage Development in Woven Fabric and Hybrid Fabric Composites. Proceedings of Sixth International Conference on Composite Materials (ICCM-VI) & Second European Conference on Composite Materials (ECCM-II), 89–99, 1987.

[24] Fujii T., Amijima S., and Okubo K., Microscopic fatigue processes in a plain-weave glass-fibre composite, Composites Science and Technology, 49, 327–333, 1993.

[25] Xiao J., Bathias C., Fatigue damage and fracture mechanism of notched woven laminates. Composite Materials, 28, 1127–1139, 1994.

[26] Lye S.W. and Boey F.Y.C., Development of a low-cost prototype filament-winding system for composite components. Materials Processing Technology, 52(2–4), 570–584, 1995.

[27] Dharan C.K.H., Fatigue failure in graphite fibre and glass fibre-polymer composites, Materials Science 10, 1665–1670, 1975.

[28] Dharan, C.K.H., The Fatigue Behavior of Fiber-Reinforced Polymers and Advanced Composites, ASME Design Engineering Conference, ASME Paper No. 77-DE-41, American Society of Mechanical Engineers (Paper), New York, 1977.

[29] Talreja R., Fatigue Damage Mechanisms, Chapter (2), Modeling Damage, Fatigue and Failure of Composite Materials, Talreja R., and Varna J., A Volume in Woodhead Publishing Series in Composites Science and Engineering, (ISBN: 978-1-78242-286-0), 2016.

[30] Duggan T.V., and Byrne J., Fatigue as a Design Criterion, McMillan Press Ltd., (ISBN: 0-333-21488-9), 1977.

[31] Kim H.C., and Ebert L.J., Axial fatigue failure sequence and mechanisms in unidirectional fiber glass composite, Composite Material, 12, 139–152, 1978.

[32] Ellyin F. and Kujawski D., Fatigue testing and life prediction of fiber glass-reinforced composites. In: Neale K.W. and Labossière P. Eds.; First International Conference on Advanced Composite Materials in Bridges and Structures (ACMBS-I), Sherbrooke, Québec, Canada, Canadian Society for Civil Engineering, 111–118, 1992.

[33] Kujawski D. and Ellyin F., Rate/frequency-dependent behaviour of fiberglass/epoxy laminates in tensile and cyclic loading, Journal of Composites, 26(10), 719–723, 1995.

[34] Lee B. L. and Liu D. S., Cumulative damage of fiber-reinforced elastomer composites under fatigue loading, Journal of Composite Materials, 28(13), 1261–1286, 1994.

[35] Chamis C. C., Mechanics of Load Transfer at the Fiber/Matrix Interface, NASA TN D-6588, February 1972.

[36] Soden P.D., Kitching R., Tse P. C., Tsavalas Y. and Hinton M. J., Influence of winding angle on the strength and deformation of filament-wound composite tubes subjected to uniaxial and biaxial loads, Journal of Composites Science and Technology, 1993, 46(4), pp. 363–378.

[37] Keck S. and Fulland M., Effect of fibre volume fraction and fibre direction on crack paths in flax fibre reinforced composites, in press, Engineering Fracture Mechanics, Available online 13 April 2016.

[38] Ghamarian N., Hanim M.A., Penjumras P. and Majid D.L., Effect of fiber orientation on the mechanical properties of laminated polymer composites, Reference Module in Materials Science and Materials Engineering, Current as of 22 July 2016.

[39] Chawla N., Liaw P.K., LaraCurzio E., Ferber M.K. and Lowden R.A., Effect of fiber fabric orientation on the flexural monotonic and fatigue behavior of 2D woven ceramic matrix composites, Materials Science and Engineering: A, 557(15),77–83, 2012.

[40] El Kadi H., and Ellyin F., Effect of stress ratio on the fatigue of unidirectional fibre glass-epoxy composite laminae, Journal of Composite Material, 25(10), 917–924, 1994.

[41] Fujii T., Shiina T. and Okubo K., Fatigue notched sensitivity of glass woven fabric composites having a circular hole under tension/torsion biaxial loading, Journal of Composite Materials, 28, 234–251, 1994.

[42] Athijayamani A., Thiruchitrambalam M., Natarajan U. and Pazhanivel B., Effect of moisture absorption on the mechanical properties of randomly oriented natural fibers/polyester hybrid composite, Journal of Materials Science and Engineering A, 2009, 517, pp.344–353.

[43] Merah N., Nizamuddin S., Khan Z. and Al-Sulaiman F, Effects of harsh weather and seawater on glass fiber reinforced epoxy composite, Journal of Reinforced Plastics and Composites, 2010, 29(20), pp.3104–3110.

[44] Kaynak C. and Mat O., Uniaxial fatigue behavior of filament-wound glass-fiber/epoxy composite tubes, Journal of Composites Science and Technology, 2001, 61(13), pp. 1833–1840.

[45] Wisnom M.R., Size effects in composites, reference module in materials science and materials engineering, Current as of 28 October 2015.

[46] Duggan T. V. and Byrne J., Fatigue as a Design Criterion, McMillan Press Ltd., 1977 (ISBN: 0-333-21488-9).

[47] Subramanian S., Elmore I.S., Stinchcomb W.W. and Reifsnider K.L., Influence of Fiber–Matrix Interphase on the long-Term Behavior of graphite/Epoxy Composite, In: Deo R.B., Saff C.R., Eds.; Composite Material: Testing and Design. ASTM STP 1274, American Society for Testing and Materials, 12, 69–87, 1996.

[48] Hochard Ch., Miot St. and Thollon Y., Fatigue of laminated composite structures with stress concentrations, Composites Part B: Engineering, 65, 11–16, 2014.

[49] Sims D.F., and Brogdon V.H., Fatigue Behavior of Composites under Different Loading Modes, Fatigue of Filamentary Materials, ASTM STP 636, K L Reifsnider and K N Lauraitis, Eds.; 185–205, 1977.

[50] Hahn H.T., Fatigue behavior and life prediction of composite laminates, Composite Materials: Testing and Design (Fifth Conference), ASTM STP 674, Tsai S.W., Ed., 383–417, 1979.

[51] Hashin Z., Fatigue failure criteria for unidirectional fiber composites, Applied Mechanics, 48, 846–852, 1981.

[52] Hashin Z., Fatigue failure criteria for combined cyclic stress, Int. J. Fracture, 17(2), 101–109, 1981.

[53] Tennyson R.C., Hansen J.S., Heppler G.R., Mabson G., Wharram G., and Street K.N., Computation of influence of defects on static and fatigue strength of composites, AGARD-CP- 355, 14–17, 1983.

[54] Wu C.M.L., Thermal and mechanical fatigue analysis of angle-ply CFRP laminates, Second International Composites Conference and Exhibition, Ottawa, Ontario, Canada, 631–638, 1993.

[55] Altabey, W. A. and Noori, M. (2018). Fatigue life prediction for carbon fibre/epoxy laminate composites under spectrum loading using two different neural network architectures. International Journal of Sustainable Materials and Structural Systems (IJSMSS), 3(1), http://dx.doi.org/10.1504/IJSMSS.2017.10013394

[56] Zhao, Y., Noori, M., Altabey, W. A., Ramin, G. and Wu, Z. (2019). A fatigue damage model for FRP composite laminate systems based on stiffness reduction. Structural Durability and Health Monitoring, 13(1), 85–103, http://dx.doi.org/10.32604/sdhm.2019.04695.

[57] Hanh H. T. and Tsai S. W., On the behaviour of composite laminates after initial failures, Composite Materials, 8, 280–305, 1974.

[58] Azzi V. D. and Tsai S. W., Anisotropic strength of composites – Investigation aimed at developing a theory applicable to laminated as well as unidirectional composites, employing simple material properties derived from unidirectional specimens alone, Experimental Mechanics, 5(9), 283–288, 1965.

[59] Puck A. and Schneider W., On failure mechanisms and failure criteria of filament wound glass fibre/resin composites, Plastics & Polymers, 37, 270–273, 1969.

[60] Sim D. F. and Brogdon V. H., Fatigue behaviour of composites under different loading modes. Fatigue of Filamentary Materials, ASTM STP 636, 185–205, 1977.

[61] Hinton M. J. and Soden P. D., Predicting failure in composite laminates: The background to the exercise, Composites Science and Technology, 58(7), 1001–1010, 1998.

[62] Norris, C.B., Strength of orthotropic materials subjected to combined stresses, Report 1816, Forest Product Laboratory, 1962.

[63] Hoffman, O., The brittle strength of orthotropic materials, Composite Materials, 1, 200–206, 1967.

[64] Pal, P., Ray, C., Progressive failure analysis of laminated composite plates by finite element method, Reinforced Plastics and Composites, 21, 1505–1513, 2002.

[65] Reddy, J.N., Pandey, A.K, A first-ply failure analysis of composite laminates, Computers and Structures, 25, 371–393, 1987.

[66] Marin, J., Theories of strength for combined stresses and nonisotropic materials, Aeronautical Sciences, 24, 265–268, 1957.

[67] Tsai S. W. and Wu E. M., A general theory of strength for anisotropic materials, Composite Materials, 5, 58–80, 1971.

[68] Philippidis T. P. and Vassilopoudos A. P., Fatigue strength prediction under multiaxial stress, Composite Materials, 33(17), 1578–1599, 1999.

[69] Hashin Z., Failure criteria for unidirectional fibre composites, Applied Mechanics, 47, 329–334, 1980.

[70] Rotem A., Prediction of laminate failure with the Rotem failure criterion, Composites Science and Technology, 58(7), 1083–1094, 1998.

[71] Ashkenazi, E. K., Problems of the anisotropy of strength, Polymer Mechanics, 1, 60–70, 1965.

[72] Guess, T.R., Biaxial testing of composite cylinders: Experimental theoretical comparison, Composites, 11, 139–148, 1980.

[73] Theocaris, P.S., A simple biaxial test for exploring failure tensor polynomial criteria of composites, Composites Science and Technology, 49, 237–249, 1993.

[74] Gargiulo, C., Marchetti, M., Rizzo, A., Prediction of failure envelopes of composite tubes subjected to biaxial loadings, Acta Astronautica, 39(5), 355–368, 1996.

[75] Cowin, S.C., Fabric dependence of an anisotropic strength criterion, Mechanics of Materials 5, 251–260, 1986.

[76] Fischer, L., Optimization of orthotropic laminates, Engineering for Industry, 89, 399–402, 1967.

[77] Rytter A., Vibration Based Inspection of Civil Engineering Structures, PhD, Aalborg University, Denmark, 1993.

[78] Amaro A.M., Santos J.B., and Cirne J.S., Delamination Depth in Composites Laminates with Interface Elements and Ultrasound Analysis, Strain, 47, 138–145, 2011.

[79] ZengHua L., HongTao Y., CunFu H., and Bin W., Delamination damage detection of laminated composite beams using air-coupled ultrasonic transducers, Physics, Mechanics & Astronomy, 56 (7), 1269–1279, 2013.

[80] Alem B., and Abedian A., Fatigue Damage Detection in Large Thin Wall Plate Based on Ultrasonic Guided Wave by Using a Piezoelectric Sensor Network, 29th Congress of the International Council of the Aeronautical Sciences, St. Petersburg, Russia, September 7–12, 2014.

[81] Liu Z., Yu H., He C., and Wu B., Delamination detection in composite beams using pure Lamb mode generated by air-coupled ultrasonic transducer, Intelligent Material Systems and Structures, 25(5), 541–550, 2014.

[82] Park B., An Y., and Sohn H., Visualization of hidden delamination and debonding in composites through noncontact laser ultrasonic scanning, Composites Science and Technology, 100 (21), 10–18, 2014.

[83] Kersemans M., Martens A., Degrieck J., Abeele K.V.D., Delrue S., Pyl L., Zastavnik F., Sol H. and Paepegem W.V., The ultrasonic polar scan for composite characterization and damage assessment: past, present and future, Applied Sciences, 58(6), 1–15, 2015.

[84] Lissenden C.J., Liu Y., and Rose J.L., Use of non-linear ultrasonic guided waves for early damage detection, Insight – Non-Destructive Testing and Condition Monitoring, 57(4) 2015.

[85] De Albuquerque V.C., Tavares J.R.S., and Durão L.M.P., Evaluation of delamination damage on composite plates using an artificial neural network for the radiographic image analysis, Composite Materials, 44 (9), 1139–1159, 2010.

[86] Tompson C.G., and Johnson W.S., Determination of the nontraditional lay-up influence and loading configuration on fatigue damage development under bearing-bypass loading conditions using radiography, Composite Materials, 45(22), 2259–2269, 2011.

[87] Aidi B., Philen M.K., and Case S.W., Progressive damage assessment of centrally notched composite specimens in fatigue, Composites: Part A, 74, 47–59, 2015.

[88] Jespersen K.M., Lowe T., Withers P.J., Zangenberg J., and Mikkelsen L.P., Micromechanical Time-Lapse X-ray CT Study of Fatigue Damage in Uni-Directional Fibre Composites, 20th International Conference on Composite Materials Copenhagen, 19–24 July 2015.

[89] Szwedo M., Bednarz J., Paćko P., Pieczonka Ł., and Uhl T., Approach to Thermographical Damage Detection in Composite Plates, Chapter, Selected Problems of Modal Analysis of Mechanical Systems, Tadeusz U., Publishing House of the Institute for Sustainable Technologies – National Research Institute (ITeE-PIB), 2009.

[90] Kêdziora P., Detection of interlinear cracks in composite structures with the use of piezoelectric sensors and thermography, III ECCOMAS Thematic Conference on Computational Methods in Structural Dynamics and Earthquake Engineering, Corfu, Greece, 25–28 May 2011.

[91] Colombo C., Libonati F. and Vergani L., Fatigue damage in GFRP, Int. J. Structural Integrity, 3 (4), 424–440, 2012.

[92] Toscano C., Riccio A., Camerlingo F., and Meola C., Lock in thermography to monitor propagation of delamination in CFRP composites during compression tests, 11th International Conference on Quantitative InfraRed Thermography, Naples, Italy, 11–14 June 2012.

[93] Schmutzler H., Alder M., Kosmann N., Wittich H., and Schulte K., Degradation monitoring of impact damaged carbon fibre reinforced polymers under fatigue loading with pulse phase thermography, Composites: Part B, 59, 221–229, 2014.

[94] Davijani A.A.B., Hajikhani M., and Ahmadi M., Acoustic emission based on sentry function to monitor the initiation of delamination in composite materials, Materials and Design, 32, 3059–3065, 2011.

[95] Jr F.A., Ozevin D., Awerbuch J. and Tan T., Detecting and locating damage initiation and progression in full-scale sandwich composite fuselage panels using acoustic emission, Composite Materials, 0(0), 1–22, 2012.

[96] Assarar M., Bentahar M., El Mahi A. and El Guerjouma R., Monitoring of damage mechanisms in sandwich composite materials using acoustic emission, Int. J. Damage Mechanics, 0(0), 1–18, 2014.

[97] Saeedifar M., Fotouhi M., Najafabadi M.A., and Toudeshky H.H., Prediction of delamination growth in laminated composites using acoustic emission and Cohesive Zone Modeling techniques, Composite Structures, 124, 120–127, 2015.

[98] Saeedifar M., Fotouhi M., Najafabadi M.A., Toudeshky H.H., and Minak G., Prediction of quasi-static delamination onset and growth in laminated composites by acoustic emission, Composites Part B: Engineering, 85, 113–122, 2016.

[99] Lakhdar M., Mohammed D., Boudjemâa L., Rabiâ A., and Bachir M., Damages detection in a composite structure by vibration analysis, TerraGreen 13 International Conference 2013 – Advancements in Renewable Energy and Clean Environment, Energy Procedia, 36, 888–897, 2013.

[100] Waghulde K.B., and Kumar B., Vibration analysis for damage detection in composite plate by using piezoelectric sensors, Int. J. Mechanical Engineering and Technology (IJMET), 5 (12), 27–35, 2014.

[101] Garcia D., Palazzetti R., Trendafilova I., Fiorini C., and Zucchelli A., Vibration-based delamination diagnosis and modelling for composite laminate plates, Composite Structures, 130, 155–162, 2015.

[102] Habtour E., Cole D.P., Riddick J.C., Weiss V., Robeson M., Sridharan R., and Dasgupta A., Detection of fatigue damage precursor using a nonlinear vibration approach, Structure Control Health Monitoring, DOI: 10.1002/stc.1844, 2016.

[103] Altabey, W.A. (2017). Free vibration of basalt fiber reinforced polymer (FRP) laminated variable thickness plates with intermediate elastic support using finite strip transition matrix (FSTM) method. Vibroengineering, 19(4), 2873–2885, https://doi.org/10.21595/jve.2017.18154.

[104] Altabey, W.A. (2017). Prediction of natural frequency of basalt fiber reinforced polymer (FRP) laminated variable thickness plates with intermediate elastic support using artificial neural networks (ANNs) method. Vibroengineering, 19(5), 3668–3678, https://doi.org/10.21595/jve.2017.18209.

[105] Altabey, W.A. (2018). High performance estimations of natural frequency of basalt FRP laminated plates with intermediate elastic support using response surfaces method. Vibroengineering, 20(2), 1099–1107, https://doi.org/10.21595/jve.2017.18456.

[106] Al-tabey, W.A. (2014). Vibration Analysis of Laminated Composite Variable Thickness Plate Using Finite Strip Transition Matrix Technique, In MATLAB Verifications MATLAB-Particular for Engineer; Kelly, B., Ed.; InTech, USA, Volume 21, pp.583–620 (2014). ISBN 980-953-307-1128-8, https://doi.org/10.5772/57384.

[107] Zhao Y., Noori M., Altabey W. A. (2020). Reaching Law Based Sliding Mode Control for a Frame Structure under Seismic Load, In Press, Earthquake Engineering and Engineering Vibration, http://https://www.springer.com/journal/11803.

[108] Heuer H., Schulze M.H. and Meyendorf N., Non-destructive evaluation (NDE) of composites: Eddy current techniques, Chapter (3), Non-Destructive Evaluation (NDE) of Polymer Matrix Composites, (ISBN: 978-0-85709-344-8), 2013.

[109] Kuang K., and Cantwell W., Use of conventional optical fibers and fiber Bragg gratings for damage detection in advanced composite structures: A review, American Society of Mechanical Engineers, Applied Mechanics Review, 56 (5), 493–513, 2003.

[110] Takeda N., Okabe Y., and Mizutani T., Damage detection in composites using optical fibre sensors, Aerospace Engineering Part G, 221, 221: 497, 2007.

[111] Peng Q., Zhang X., Huang C., Carter E.A., and Lu G., Hierarchical fiber-optic delamination detection system for carbon fiber reinforced plastic structures, Modelling Simulation Materials Science and Engineering, 18, 1–14, 2012.

[112] Zuluaga-Ramírez P., Arconada Á., Frövel M., Belenguer T., and Salazar F., Optical sensing of the fatigue damage state of CFRP under realistic aeronautical load sequences, Sensors, 15, 5710–5721, 2015.

[113] Altabey, W.A., Noori, Alarjani, A. and Zhao, Y. (2020). Nano-delamination monitoring of BFRP nano-pipes of electrical potential change with ANNs. Advances in Nano Research, 9(1), 1–13, http://dx.doi.org/10.12989/anr.2020.9.1.001.

[114] Zhao, Y., Noori, M., Altabey, W.A., and Wu, Z. (2018). Fatigue damage identification for composite pipeline systems using electrical capacitance sensors. Smart Materials and Structures, 27(8), 085023, https://doi.org/10.1088/1361-665X/aacc99.

[115] Wang T., Noori M., Altabey W. A. (2020). Identification of cracks in an Euler-Bernoulli beam using Bayesian inference and closed-form solution of vibration modes, In Press, Part L: Journal of Materials: Design and Applications, https://journals.sagepub.com/home/pila.

[116] Li Z., Noori M., Altabey W. A. (2020). An experimental study on the seismic performance of adobe walls, In Press, Structural Durability and Health Monitoring, http://www.tspsubmission.com/index.php/sdhm.

[117] Altabey, W.A. (2017). Delamination evaluation on basalt FRP composite pipe of electrical potential change. Advances in Aircraft and Spacecraft Science, 4(5), 515–528, http://dx.doi.org/10.12989/aas.2017.4.5.515.

[118] Altabey, W.A. (2017). EPC method for delamination assessment of basalt FRP pipe: Electrodes number effect. Structural Monitoring and Maintenance, 4(1), 69–84, https://doi.org/10.12989/smm. 2017.4.1.069.

[119] Altabey, W.A. and Noori, M. (2017). Detection of fatigue crack in basalt FRP laminate composite pipe using electrical potential change method. 12th International Conference on Damage Assessment of Structures, IOP Conf. Series: Journal of Physics, 842, 012079.

[120] Altabey, W. A., Noori, M., Wu, Z., Al-Moghazy, M. A., Kouritem, S. A. (2023), A deep-learning approach for predicting water absorption in composite pipes by extracting the material's dielectric features, Engineering Applications of Artificial Intelligence, 121, 105963, https://doi.org/10.1016/j.en gappai.2023.105963.

[121] Altabey, W.A., A comprehensive study of a long-term creep thermo-mechanical fatigue behavior monitoring of BFRP composite pipeline using electrical capacitance sensors and deep learning algorithm, International Journal of Fatigue, 2024,. https://doi.org/10.1016/j.ijfatigue.2024.108277.

[122] Altabey, W.A. (2016). FE and ANN model of ECS to simulate the pipelines suffer from internal corrosion. Structural Monitoring and Maintenance, 3(3), 297–314, http://dx.doi.org/10.12989/smm. 2016.3.3.297.

[123] Altabey, W.A. (2016). Detecting and Predicting the Crude Oil type inside Composite Pipes Using ECS and ANN. Structural Monitoring and Maintenance, 3(4), 377–393, http://dx.doi.org/10.12989/smm. 2016.3.4.377.

[124] Altabey, W.A. (2016). The thermal effect on electrical capacitance sensor for two-phase flow monitoring. Structural Monitoring and Maintenance, 3(4), 335–347, http://dx.doi.org/10.12989/ smm.2016.3.4.335.

[125] Dayal V., and Kinra V.K., Leaky Lamb waves in an anisotropic plate. II- Nondestructive evaluation of matrix cracks in fiber-reinforced composites, Acoustic Society, 89 (4), 1590–1598, 1991.

[126] Kessler S.S., Spearing S.M. and Soutis C., Damage detection in composite materials using Lamb wave methods, Smart Material Structure, 11, 269–278, 2002.

[127] Yashiro S., Takatsubo J., and Toyama N., An NDT technique for composite structures using visualized Lamb-wave propagation, Composites Science and Technology, 67, 3202–3208, 2007.

[128] Hu N., Liu Y., Li Y., Peng X., and Yan B., Optimal excitation frequency of lamb waves for delamination detection in CFRP laminates, Composite Materials, 44, (13), 2010.

[129] Yeum C.M., Sohn H., Ihn J.B., and H.J., Delamination detection in a composite plate using a dual piezoelectric transducer network, Composite Structures, 94, 3490–3499, 2012.

[130] Gopalakrishnan S., Lamb wave propagation in laminated composite structures, The Indian Institute of Science, 93 (4), 699–713, 2013.

[131] Yeum C.M., Sohn H., Lim H.J., and Ihn J.B., Reference-free delamination detection using Lamb waves, Structure Control Health Monitoring, 21, 675–684, 2014.

[132] Keulen C.J., Yildiz M., and Suleman A., Damage detection of composite plates by Lamb wave ultrasonic tomography with a sparse hexagonal network using damage progression trends, Shock and Vibration, 1–8, 2014.

[133] Spiegel M.D., Damage Detection in Composite Materials Using PZT Actuators and Sensors for Structural Health Monitoring, Master, Department of Electrical and Computer Engineering, University of Alabama, 2014.

[134] Qiao P., and Fan W., Lamb wave-based damage imaging method for damage detection of rectangular composite plates, Structural Monitoring and Maintenance,1 (4), 411–425, 2014.

[135] Shen Y., and Giurgiutiu V., Combined analytical FEM approach for efficient simulation of Lamb wave damage detection, Ultrasonics, 69, 116–128, 2016.

[136] Wang, E.; Cheng, P.; Li, J.; Cheng, Q.; Zhou, X.; Jiang, H. High-sensitivity temperature and magnetic sensor based on magnetic fluid and liquid ethanol filled micro-structured optical fiber. Opt. Fiber Technol. 2020, 55, 102161.

[137] Arun Francis, G.; Arulselvan, M.; Elangkumaran, P.; Keerthivarman, S.; Vijaya Kumar, J. Object detection using ultrasonic sensor. Int. J. Innov. Technol. Explor. Eng. 2020, 8, 207–209.

[138] Giannì, C.; Balsi, M.; Esposito, S.; Ciampa, F. Low-power global navigation satellite system-enabled wireless sensor network for acoustic emission localisation in aerospace components. Struct. Control. Health Monit. 2020, 27, e2525.

[139] Cavaliere, M.; McVeigh, O.; Jaeger, H.A.; Hinds, S.; O'Donoghue, K.; Cantillon-Murphy, P. Inductive sensor design for electromagnetic tracking in image guided interventions. IEEE Sens. J. 2020, 20, 8623–8630.

[140] Luo, B.; Long, T.; Guo, L.; Dai, R.; Mai, R.; He, Z. Analysis and design of inductive and capacitive hybrid wireless power transfer system for railway application. IEEE Trans. Ind. Appl. 2020, 56, 3034–3042.

[141] Jalal, A.; Quaid, M.A.K.; Tahir, S.B.U.D.; Kim, K. A study of accelerometer and gyroscope measurements in physical life-log activities detection systems. Sensors 2020, 20, 6670.

[142] Tan, X.; Sun, Z.; Wang, P.; Sun, Y. Environment-aware localization for wireless sensor networks using magnetic induction. Hoc Netw. 2020, 98, 102030.

[143] Hasan, M.N.; Salman, M.S.; Islam, A.; Znad, H.; Hasan, M.M. Sustainable composite sensor material for optical cadmium (II) monitoring and capturing from wastewater. Microchem. J. 2021, 161, 105800.

[144] Qing, X.; Liu, X.; Zhu, J.; Wang, Y. In-situ monitoring of liquid composite molding process using piezoelectric sensor network. Struct. Health Monit. 2020.

[145] Tay, R.Y.; Li, H.; Lin, J.; Wang, H.; Lim, J.S.K.; Chen, S.; Leong, W.L.; Tsang, S.H.; Teo, E.H.T. Lightweight, superelastic boron nitride/polydimethylsiloxane foam as air dielectric substitute for multifunctional capacitive sensor applications. Adv. Funct. Mater. 2020, 30, 1909604.

[146] Nauman, S.; Asfar, Z.; Ahmed, S.; Nasir, M.A.; Hocine, N.A. On the in-situ on-line structural health monitoring of composites using screen-printed sensors. Thermoplast. Compos. Mater. 2021.

[147] Tuloup, C.; Harizi, W.; Aboura, Z.; Meyer, Y.; Ade, B.; Khellil, K. Detection of the key steps during Liquid Resin Infusion manufacturing of a polymer-matrix composite using an in-situ piezoelectric sensor. Mater. Today Commun. 2020, 24, 101077.

[148] Georgopoulou, A.; Michel, S.; Vanderborght, B.; Clemens, F. Piezoresistive sensor fiber composites based on silicone elastomers for the monitoring of the position of a robot arm. Sens. Actuators A Phys. 2021, 318, 112433.

[149] Georgopoulou, A.; Clemens, F. Piezoresistive elastomer-based composite strain sensors and their applications. ACS Appl. Electron. Mater. 2020, 2, 1826–1842.

[150] Wang, M.; Li, N.; Wang, G.D.; Lu, S.W.; Di Zhao, Q.; Liu, X.L. High-sensitive flexural sensors for health monitoring of composite materials using embedded carbon nanotube (CNT) buckypaper. Compos. Struct. 2021, 261, 113280.

[151] Nauman, S. Piezoresistive sensing approaches for structural health monitoring of polymer composites – a review. Eng 2021, 2, 197–226.

[152] Bednarska, K.; Sobotka, P.; Woliński, T.R.; Zakrecka, O.; Pomianek,W.; Nocoń, A.; Lesiak, P. Hybrid fiber optic sensor systems in structural health monitoring in aircraft structures. Materials 2020, 13, 2249.

[153] Güemes, A.; Fernandez-Lopez, A.; Pozo, A.R.; Sierra-Pérez, J. Structural health monitoring for advanced composite structures: A review. Compos. Sci. 2020, 4, 13.

[154] Fazzi, L.; Valvano, S.; Alaimo, A.; Groves, R.M. A simultaneous dual-parameter optical fibre single sensor embedded in a glass fibre/epoxy composite. Compos. Struct. 2021, 270, 114087.

[155] Dai, H.; Thostenson, E.T.; Schumacher, T. Comparative study of the thermoresistive behavior of carbon nanotube-based nanocomposites and multiscale hybrid composites. Compos. Part B Eng. 2021, 222, 109068.

[156] Karalis, G.; Tzounis, L.; Tsirka, K.; Mytafides, C.K.; Voudouris Itskaras, A.; Liebscher, M.; Lambrou, E.; Gergidis, L.N.; Barkoula, N.M.; Paipetis, A.S. Advanced glass fiber polymer composite laminate operating as a thermoelectric generator: a structural device for micropower generation and potential large-scale thermal energy harvesting. ACS Appl. Mater. Interfaces 2021, 13, 24138–24153.

[157] Shu, Q.; Hu, T.; Xu, Z.; Zhang, J.; Fan, X.; Gong, X.; Xuan, S. Non-tensile piezoresistive sensor based on coaxial fiber with magnetoactive shell and conductive flax core. Compos. Part A Appl. Sci. Manuf. 2021, 149, 106548.

[158] Gao, F.; Shao, Y.; Hua, J.; Zeng, L.; Lin, J. Enhanced wavefield imaging method for impact damage detection in composite laminates via laser-generated Lamb waves. Measurement 2021, 173, 108639.

[159] Ding, G.; Song, W.; Gao, X.; Cao, H. Damage detection in holed carbon fiber composite laminates using embedded fiber Bragg grating sensors based on strain information. Shock Vib. 2020, 2020, 8813213.

[160] Huang, L.; Zeng, L.; Lin, J.; Zhang, N. Baseline-free damage detection in composite plates using edge-reflected Lamb waves. Compos. Struct. 2020, 247, 112423.

[161] Pérez, M.A.; Pernas-Sánchez, J.; Artero-Guerrero, J.; Serra-López, R. High-velocity ice impact damage quantification in composite laminates using a frequency domain-based correlation approach. Mech. Syst. Signal Process. 2021, 147, 107124.

[162] Shoja, S.; Berbyuk, V.; Boström, A. Delamination detection in composite laminates using low frequency guided waves: Numerical simulations. Compos. Struct. 2018, 203, 826–834.

[163] Dang, X. Statistic strategy of damage detection for composite structure using the correlation function amplitude vector. Procedia Eng. 2015, 99, 1395–1406.

[164] Zhou, J.; Li, Z.; Chen, J. Damage identification method based on continuous wavelet transform and mode shapes for composite laminates with cutouts. Compos. Struct. 2018, 191, 12–23.

[165] Yelve, N.P.; Mitra, M.; Mujumdar, P. Detection of delamination in composite laminates using Lamb wave based nonlinear method. Compos. Struct. 2017, 159, 257–266.

[166] Zhao, G.; Wang, B.; Wang, T.; Hao, W.; Luo, Y. Detection and monitoring of delamination in composite laminates using ultrasonic guided wave. Compos. Struct. 2019, 225, 111161.

[167] Mouritz A.P., Non-destructive Evaluation of Damage Accumulation, Chapter (8), Fatigue in Composites, Harris B., A Volume in Woodhead Publishing Series in Composites Science and Engineering, (ISBN: 978-1-85573-608-5), 2003.

[168] Giurgiutiu, V. Structural Health Monitoring with Piezoelectric Wafer Active Sensors, 2nd ed.; Academic Press, an Imprint of Elsevier: Amsterdam, The Netherlands, 2014; ISBN 978-0-12-418691-0.

[169] Capineri, L.; Bulletti, A. Ultrasonic guided-waves sensors and integrated structural health monitoring systems for impact detection and localization: a review. Sensors 2021, 21, 2929. https://doi.org/10.3390/s21092929

[170] Safri, S.; Sultan, M.T.H.; Yidris, N.; Mustapha, F. Low velocity and high velocity impact test on composite materials – a review. Int. J. Eng. Sci. 2014, 3, 50–60.

[171] Ziola, S.M.; Gorman, M.R. Source location in thin plates using cross-correlation. Acoust. Soc. Am. 1991, 90, 2551–2556.

[172] Marino-Merlo, E.; Bulletti, A.; Giannelli, P.; Calzolai, M.; Capineri, L. Analysis of errors in the estimation of impact positions in plate-like structure through the triangulation formula by piezoelectric sensors monitoring. Sensors 2018, 18, 3426.

[173] Ross, R. Structural Health Monitoring and Impact Detection Using Neural Networks for Damage Characterization. In Proceedings of the 47th AIAA/ASME/ASCE/AHS/ASC Structures, Structural Dynamics, and Materials, Newport, RI, USA, 1–4 May 2006; Volume 9.

[174] Gorman, M.R.; Humes, D.H.; June, R.; Prosser, W.H.; Prosser, W.H. Acoustic emission signals in thin plates produced by impact damage. Acoust. Emiss. 1999, 17, 29–36.

[175] Yang, J.C.S.; Chun, D.S. Application of the Hertz Contact Law to Problems of Impact in Plates; Defense Technical Information Center: Fort Belvoir, WV, USA, 1969.

[176] Na, W.; Baek, J. Impedance-based non-destructive testing method combined with unmanned aerial vehicle for structural health monitoring of civil infrastructures. Appl. Sci. 2016, 7, 15.

[177] Yan, W.; Chen, W.Q. Structural health monitoring using high-frequency electromechanical impedance signatures. Adv. Civ. Eng. 2010, 2010, 1–11.

[178] Annamdas, V.G.; Radhika, M.A. Electromechanical impedance of piezoelectric transducers for monitoring metallic and non-metallic structures: A review of wired, wireless and energy-harvesting methods. Intell. Mater. Syst. Struct. 2013, 24, 1021–1042.

[179] Baptista, F.; Budoya, D.; Almeida, V.; Ulson, J. An experimental study on the effect of temperature on piezoelectric sensors for impedance-based structural health monitoring. Sensors 2014, 14, 1208–1227.

[180] Ai, D.; Zhu, H.; Luo, H.; Wang, C. Mechanical impedance based embedded piezoelectric transducer for reinforced concrete structural impact damage detection: A comparative study. Constr. Build. Mater. 2018, 165, 472–483.

[181] Park, G.; Cudney, H.H.; Inman, D.J. An integrated health monitoring technique using structural impedance sensors. Intell. Mater. Syst. Struct. 2000, 11, 448–455.

[182] Overly, T.G.S.; Park, G.; Farinholt, K.M.; Farrar, C.R. Development of an extremely compact impedance-based wireless sensing device. Smart Mater. Struct. 2008, 17, 065011.

[183] Park, S.; Ahmad, S.; Yun, C.-B.; Roh, Y. Multiple crack detection of concrete structures using impedance-based structural health monitoring techniques. Exp. Mech. 2006, 46, 609–618.

[184] Lim, Y.Y.; Bhalla, S.; Soh, C.K. Structural identification and damage diagnosis using self-sensing piezo-impedance transducers. Smart Mater. Struct. 2006, 15, 987–995.

[185] Na, W.; Baek, J. A review of the piezoelectric electromechanical impedance based structural health monitoring technique for engineering structures. Sensors 2018, 18, 1307.

[186] Liang, C.; Sun, F.; Rogers, C. An impedance method for dynamic analysis of active material systems. In Proceedings of the 34th Structures, Structural Dynamics and Materials Conference, La Jolla, CA, USA, 19–22 April 1993; Volume 116, pp. 120–128.

[187] Liang, C.; Sun, F.P.; Rogers, C.A. Coupled electro-mechanical analysis of adaptive material systems-determination of the actuator power consumption and system energy transfer. Intell. Mater. Syst. Struct. 1994, 5, 12–20.

[188] Park, G.; Inman, D.J. Structural health monitoring using piezoelectric impedance measurements. Philos. Trans. R. Soc. A Math. Phys. Eng. Sci. 2006, 365, 373–392.

[189] Park, G.; Sohn, H.; Farrar, C.R.; Inman, D.J. Overview of piezoelectric impedance-based health monitoring and path forward. Shock Vib. Dig. 2003, 35, 451–463.

[190] Thien, A.; Chiamori, H.; Ching, J.;Wait, J.; Park, G. Piezoelectric active sensing for damage detection in pipeline structures. In Proceedings of the 23rd International Modal Analysis Conference, Orlando, FL, USA, 31 January–5 February 2005; pp. 323–636.

[191] Baptista, F.G.; Filho, J.V. A new impedance measurement system for PZT-based structural health monitoring. IEEE Trans. Instrum. Meas. 2009, 58, 3602–3608.

[192] Park, S.; Park, G.; Yun, C.-B.; Farrar, C.R. Sensor self-diagnosis using a modified impedance model for active sensing-based structural health monitoring. Struct. Health Monit. 2008, 8, 71–82.

[193] Park, S.; Yun, C.-B.; Roh, Y.; Lee, J.-J. PZT-based active damage detection techniques for steel bridge components. Smart Mater. Struct. 2006, 15, 957–966.

[194] Na, W.; Baek, J. Adhesive defect monitoring of glass fiber epoxy plate using an impedance-based non-destructive testing method for multiple structures. Sensors 2017, 17, 1439.

[195] Giurgiutiu, V.; Zagrai, A.N. Characterization of piezoelectric wafer active sensors. Intell. Mater. Syst. Struct. 2000, 11, 959–976.

[196] Giurgiutiu, V.; Zagrai, A.; Jing Bao, J. Piezoelectric wafer embedded active sensors for aging aircraft structural health monitoring. Struct. Health Monit. 2002, 1, 41–61.

[197] Cuc, A.; Giurgiutiu, V.; Joshi, S.; Tidwell, Z. Structural health monitoring with piezoelectric wafer active sensors for space applications. AIAA J. 2007, 45, 2838–2850.

[198] Rathod, V.T.; Mahapatra, D.R.; Gopalakrishnan, S. Lamb wave based identification and parameter estimation of corrosion in metallic plate structure using a circular PWAS array. Health Monit. Struct. Biol. Syst. 2009, 2009, 7295–72951C.

[199] Rathod, V.T.; Mahapatra, D.R. Lamb wave based monitoring of plate-stiffener deboding using a circular array of piezoelectric sensors. Int. J. Smart Sens. Intell. Syst. 2010, 3, 27–44.

[200] Raghavan, A.; Cesnik, C.E.S. Modeling of piezoelectric-based Lamb wave generation and sensing for structural health monitoring. Smart Struct. Mater. 2004, 5391, 419–430.

[201] Giurgiutiu, V. Tuned Lamb wave excitation and detection with piezoelectric wafer active sensors for structural health monitoring. Intell. Mater. Syst. Struct. 2005, 16, 291–305.

[202] Giurgiutiu, V. Lamb wave generation with piezoelectric wafer active sensors for structural health monitoring. Opt. Microsys. 2003, 5056, 111–122.

[203] Wait, J.R.; Park, G.; Worden, C.R. Integrated structural health assessment using piezoelectric active sensors. Shock Vib. 2005, 12, 389–405.

[204] Park, S.; Kim, J.-W.; Lee, C.; Park, S.-K. Impedance-based wireless debonding condition monitoring of CFRP laminated concrete structures. NDT E Int. 2011, 44, 232–238.

[205] Giurgiutiu, V. Structural Health Monitoring with Piezoelectric Wafer Active Sensors; Elsevier: Amsterdam, The Netherlands, 2014. Available online: https://www.elsevier.com/books/structural-health-monitoringwith-piezoelectric-wafer-active-sensor/giurgiutiu/978-0-12-418691-0 (accessed on 30 August 2022).

[206] Giurgiutiu, V. Structural health monitoring with piezoelectric wafer active sensors. In Proceedings of the 16th International Conference of Adaptive Structures and Technologies ICAST, Paris, France, 10–12 October 2005.

[207] Gresil, M.; Giurgiutiu, V. Guided wave propagation in composite laminates using piezoelectric wafer active sensors. Aeronaut. J. 2013, 117, 971–995.

[208] Gresil, M.; Yu, L.; Giurgiutiu, V. Fatigue Crack Detection in Thick Steel Structures with Piezoelectric Wafer Active Sensors. In Nondestructive Characterization for Composite Materials Aerospace Engineering, Civil Infrastructure, and Homeland Security 2011; SPIE: Bellingham, WA, USA, 2011; p. 79832Y.

[209] Yu, L.; Giurgiutiu, V.; Pollock, P. A Multi-mode Sensing System for Corrosion Detection Using Piezoelectric Wafer Active Sensors. In Sensors and Smart Structures Technologies for Civil 2008 Mechanical, and Aerospace Systems 2008; SPIE: Bellingham, WA, USA, 2008; p. 6932.

[210] Yu, L.; Cheng, L.; Su, Z. Correlative sensor array and its applications to identification of damage in plate-like structures. Struct. Control Health Monit. 2011, 19, 650–671.

[211] Giurgiutiu, V.; Bao, J.; Zhao, W. Piezoelectric wafer active sensor embedded ultrasonics in beams and plates. Exp. Mech. 2003, 43, 428–449.

[212] Qing, X.P.; Beard, S.J.; Ikegami, R.; Chang, F.-K.; Boller, C. Aerospace applications of SMART layer technology. In Encyclopedia of Structural Health Monitoring; Boller, C., Chang, F.-K., Fujino, Y., Eds.; Wiley: Chichester, UK, 2009.

[213] Lin, M.; Qing, X.; Kumar, A.; Beard, S.J. SMART layer and SMART suitcase for structural health monitoring applications. Int. Soc. Opt. Photonics 2001, 4332, 98–106.

[214] Qing, X.P.; Beard, S.J.; Kumar, A.; Ooi, T.K.; Chang, F.-K. Built-in sensor network for structural health monitoring of composite structure. Intell. Mater. Syst. Struct. 2006, 18, 39–49.

[215] Qing, X.P.; Beard, S.J.; Kumar, A.; Li, I.; Lin, M.; Chang, F.-K. Stanford multiactuator–receiver transduction (SMART) layer technology and its applications. In Encyclopedia of Structural Health Monitoring; Boller, C., Chang, F.-K., Fujino, Y., Eds.; Wiley: Chichester, UK, 2009.

[216] Haywood, J.; Coverley, P.T.; Staszewski, W.J.; Worden, K. An automatic impact monitor for a composite panel employing smart sensor technology. Smart Mater. Struct. 2004, 14, 265–271.

[217] Qing, X.P.; Chan, H.-L.; Beard, S.J.; Ooi, T.K.; Marotta, S.A. Effect of adhesive on the performance of piezoelectric elements used to monitor structural health. Int. J. Adhes. Adhes. 2006, 26, 622–628.

[218] Park, J.; Chang, F.-K. System identification method for monitoring impact events. Int. Soc. Opt. Photonics 2005, 5758, 189–200.

[219] Kumar, A.; Wu, H.F.; Lin, M.; Beard, S.; Qing, X.; Zhang, C.; Hamilton, M.; Ikegami, R. Potential applications of SMART Layer technology for homeland security. Int. Soc. Opt. Photonics 2004, 5395, 61–69.

[220] Qing, X.P.; Chan, H.-L.; Beard, S.J.; Kumar, A. An active diagnostic system for structural health monitoring of rocket engines. Intell. Mater. Syst. Struct. 2006, 17, 619–628.

[221] Alavi, A.H.; Hasni, H.; Lajnef, N.; Chatti, K.; Faridazar, F. An intelligent structural damage detection approach based on self-powered wireless sensor data. Autom. Constr. 2016, 62, 24–44.

[222] Lajnef, N.; Rhimi, M.; Chatti, K.; Mhamdi, L.; Faridazar, F. Toward an integrated smart sensing system and data interpretation techniques for pavement fatigue monitoring. Comput. Aided Civ. Infrastruct. Eng. 2011, 26, 513–523.

[223] Hasni, H.; Alavi, A.H.; Jiao, P.; Lajnef, N. Detection of fatigue cracking in steel bridge girders: A support vector machine approach. Arch. Civ. Mech. Eng. 2017, 17, 609–622.

[224] Hasni, H.; Jiao, P.; Lajnef, N.; Alavi, A.H. Damage localization and quantification in gusset plates: A battery-free sensing approach. Struct. Control Health Monit. 2018, 25, e2158.

[225] Alavi, A.H.; Hasni, H.; Lajnef, N.; Chatti, K.; Faridazar, F. Damage detection using self-powered wireless sensor data: An evolutionary approach. Measurement 2016, 82, 254–283.

[226] Su Z., Ye L., and Lu Y., Guided Lamb waves for identification of damage in composite structures: A review, Sound and Vibration, 295, 753–780, 2006.

[227] Chimenti D.E., Guided waves in plates and their use in materials characterization, Applied Mechanics Reviews, 50, 47–284, 1997.

[228] Worlton D.C., Experimental confirmation of Lamb waves at megacycle frequencies, Applied Physics, 32, 967–971, 1961.

[229] Saravanos D.A. and Birman V., Detection of delaminations in composite beams using piezoelectric sensors, Proceedings of the 35th Structures, Structural Dynamics and Materials Conference of the AIAA, 1994.

[230] Percival W.J. and Birt E.A., A study of Lamb wave propagation in carbon-fibre composites, Non Destructive Testing and Condition Monitoring 39, 728–735, 1997.

[231] Seale M.D. and Smith B.T., Lamb wave assessment of fatigue and thermal damage in composites, Acoustic Society of America 103, 2416–2424, 1998.

[232] Tang B. and Henneke E.G., Lamb wave monitoring of axial stiffness reduction of laminated composite plates, Materials Evaluation, 47, 928–934, 1989.

[233] Alleyne D.N., and Cawley P., The interaction of Lamb waves with defects, IEEE Transactions on Ultrasonics, Ferroelectrics and Frequency Control, 39(3), 381–397, 1992.

[234] Saravanos D.A., Birman V., and Hopkins D.A., Detection of delaminations in composite beams using piezoelectric sensors, Proceedings of the 31st AIAA/ASME/ASCE/AHS/ASC Structures, Structural Dynamics and Materials Conference, 181–191, 1994.

[235] Kessler S.S., Spearing S.M., and Atalla M.J., In-situ damage detection of composites structures using Lamb waves methods, Proceedings of the 1st European Workshop on Structural Health Monitoring, Ecole Noemale Supérieure, Cachan, Paris, France, 374–381, 2002.

[236] Su Z., Ye L., and Bu X., Evaluation of delamination in laminated composites based on Lamb waves methods: FEM simulation and experimental verification, Proceedings of the 1st European Workshop on Structural Health Monitoring, Ecole Noemale Supérieure, Cachan, Paris, France, 328–335, 2002.

[237] Lee B.C., and Staszewski W.J., Modelling of acousto-ultrasonic wave interaction with defects in metallic structures, Proceedings of the International Conference on Noise and Vibration Engineering (ISMA 2002), Leuven, Belgium, 319–327, 2002.

[238] Ricci F., Banerjee S., and Mal A.K., Health monitoring of composite structures using wave propagation data, Proceedings of the 2nd European Workshop on Structural Health Monitoring, Forum am Deutschen Museum, Munich, Germany, 1035–1042, 2004.

[239] Toyama N., and Okabe T., Effects of tensile strain and transverse cracks on Lamb-wave velocity in cross-ply FRP laminates, Materials Science, 39, 7365–7367, 2004.

[240] Sundararaman S., Adams, D.E., and Rigas E.J., Characterizing damage in plates through beamforming with sensor arrays, Proceedings of the 23rd International Modal Analysis Conference (IMAC XXIII), Orlando, Florida, USA, paper no. 249, 2005.

[241] Kim B.H., Stubbs N., and Park T., Flexural damage index equations of a plate, Sound and Vibration, 283, 341–368, 2005.

[242] Harri K., Guillaume P., and Vanlanduit S., On-line damage detection on a wing panel using transmission of multisine ultrasonic waves, NDT & E International, 41(4), 312–317, 2008.

[243] Hurlebaus S., Niethammer M., and Jacobs L.J., Automated methodology to locate notches with lamb waves, Acoustics Research Letters Online, Acoustical Society of America, 2(4), 97–102, 2001.

[244] Jeong H., Analysis of plate wave propagation in anisotropic laminates using a wavelet transform, NDT&E International, 34, 185–190, 2001.

[245] Zhao, Y., Noori, M. and Altabey, W.A. (2017). Damage detection for a beam under transient excitation via three different algorithms. Structural Engineering and Mechanics, 64(6), 803–817, https://doi.org/10.12989/sem.2017.64.6.803.

[246] Zhao, Y., Noori M., Seyed, B. B. and Altabey, W.A. (2018). Mode shape based damage identification for a reinforced concrete beam using wavelet coefficient differences and multi-resolution analysis. Structural Control & Health Monitoring, 25(1), e2041, https://doi.org/10.1002/stc.2041.

[247] Zhao, Y., Noori, M., Altabey, W.A. and Awad, T. (2019). A comparison of three different methods for the identification of hysterically degrading structures using BWBN model. Front. Built Environ. 4(80), 1–19. http://dx.doi.org/10.3389/fbuil.2018.00080.

[248] Noori, M., Wang, H., Altabey, W.A. and Silik, A.I.H. (2018). A modified wavelet energy rate based damage identification method for steel bridges. Scientia Iranica, International Journal of Science and Technology, Transactions on Mechanical Engineering (B), 25(6), 3210–3230, https://doi.org/10.24200/sci.2018.20736.

[249] Silik A., Noori M., Altabey W. A., Ghiasi R. (2020). Comparative analysis of wavelet transform for time-frequency analysis and transient localization in structural health monitoring, In Press, Structural Durability and Health Monitoring, http://www.tspsubmission.com/index.php/sdhm.

[250] Silik A., Noori M., Altabey W. A., Ghiasi R. (2020). Choosing optimum levels of wavelet multi-resolution analysis for time-varying signals in structural health monitoring. In Press, Structural Control and Health Monitoring, https://onlinelibrary.wiley.com/journal/15452263.

[251] Su Z., and Ye L., Identification of Damage Using Lamb Waves: From Fundamentals to Applications, Springer-Verlag Berlin Heidelberg, 2009.

[252] Su Z., Ye L., and Lu Y., Guided Lamb waves for identification of damage in composite structures: A review, Sound and Vibration, 295,753–780, 2006.

[253] Ratassepp M. and Klauson A., Curvature effects on wave propagation in an infinite pipe, Ultragarsas, 59(2)-19–25, 2006.

[254] Chakraborty A., Encyclopedia of Structural Health Monitoring, Chapter (49), Modeling of Lamb Waves in Composite Structures, John Wiley & Sons, 923–939, 2009.

[255] Nayfeh A., and Chimenti D., Free wave propagation in plates of general anisotropic media, Applied Mechanics, 56(4), 881–886, 1989.

[256] Lowe M.J.S., Matrix techniques for modeling ultrasonic waves in multilayered media, IEEE Transactions on Ultrasonics, Ferroelectrics and Frequency Control, 42(4), 525–542, 1995.

[257] Nayfeh A.H., Wave Propagation in Layered Anisotropic Media with Applications to Composites, Elsevier Science B.V., 1995.

[258] Alleyne D.N., and Cawley P., A 2-dimensional Fourier transform method for the quantitative measurement of lamb modes, Proceedings of IEEE Ultrasonics Symposium, 2, 1143–1146, 1990.

[259] Hayward G, and Hyslop J., Determination of Lamb wave dispersion data in lossy anisotropic plates using time domain finite element analysis, part I: Theory and experimental verification, IEEE Transactions On Ultrasonics, Ferroelectrics, and Frequency Control, 53(2), 443–448, 2006.

[260] Toyama N., and Takatsubo J., Lamb wave method for quick inspection of impact-induced delamination in composite laminates, Composites Science and Technology, 64, 1293–1300, 2004.

[261] Schulz M.J., Pai P.F., and Inman D.J., Health monitoring and active control of composite structures using piezoceramic patches, Composites: Part B, 30, 713–725, 1999.

[262] Ditri J.J., and Rajana K., An experimental study of the angular dependence of lamb wave excitation amplitudes, Sound and Vibration, 204(5), 755–768, 1997.

[263] Schmidt D., Heinze C., Hillger W., Szewieczek A., Sinapius M., and Wierach P., Design of mode selective actuators for lamb wave excitation in composite plates, Proceedings of SPIE, 7984, 798409, 2011.

[264] Koh Y.L., Chiu W.K., and Rajic N., Effects of local stiffness changes and delamination on lamb wave transmission using surface-mounted piezoelectric transducers, Composite Structures, 57, 437–443, 2002.

[265] Williams R.B., Park G., Inman D.J., and Wilkie W.K., An overview of composite actuators with piezoceramic fibers, Proceedings of SPIE, 4753, 2002.

[266] Akdogan E.K., Allahverdi M., and Safari A., Piezoelectric composites for sensor and actuator applications, IEEE Transactions on Ultrasonics, Ferroelectrics, and Frequency Control, 52(5), 746–775, 2005.

[267] Ostachowicz W., Soman, R. Optimization of sensor placement for structural health monitoring: A review, Structural Health Monitoring 2019, 18(3), 963–988.

[268] Sun, H and Büyüköztürk, O. Optimal sensor placement in structural health monitoring using discrete optimization. Smart Mater Struct 2015; 24(12): 125034.

[269] Gomes, G.F., Da Cunha, S.S., Da Silva Lopes Alexandrino, P. et al. Sensor placement optimization applied to laminated composite plates under vibration. Struct Multidisc Optim 2018, 58, 2099–2118. https://doi.org/10.1007/s00158-018-2024-1.

[270] Yao Y and Glisic B. Detection of steel fatigue cracks with strain sensing sheets based on large area electronics. Sensors 2015; 15(4): 8088–8108.

[271] Downey A, Hu C and Laflamme S. Optimal sensor placement within a hybrid dense sensor network using an adaptive genetic algorithm with learning gene pool. Struct Health Monit 2018; 17(3): 450–460.

[272] Lu G, Xiaojin Z, Hesheng Z, et al. Optimal placement of FBG sensors for reconstruction of flexible plate structures using modal approach. In: 34th Chinese control conference (CCC), Hangzhou, China, 28–30 July 2015, 4587–4592. New York: IEEE.

[273] Cazzulani G, Chieppi M, Colombo A, et al. Optimal Sensor Placement for Continuous Optical Fiber Sensors. In: Sensors and Smart Structures Technologies for Civil, Mechanical, and Aerospace Systems, vol. 10598, Denver, CO, 4–8 March 2018, p. 1059844. Bellingham, WA: International Society for Optics and Photonics.

[274] Venkat RS, Boller C, Ravi N, et al. Optimized actuator/sensor combinations for structural health monitoring: Simulation and experimental validation. In: International Ostachowicz et al. 23 workshop on structural health monitoring, Stanford, CA, 1–3 September 2015.

[275] Ewald V, Groves RM and Benedictus R. Transducer placement option of lamb wave SHM system for hotspot damage monitoring. Aerospace 2018; 5(2): 39.

[276] Khodaei ZS and Aliabadi M. An optimization strategy for best sensor placement for damage detection and localization in complex composite structures. In: 8th European workshop on

structural health monitoring (EWSHM 2016), Bilbao, 5–8 July 2016, pp. 5–8, https://www.ndt.net/events/EWSHM2016/app/content/Paper/383_SharifKhodaei.pdf.

[277] Thiene M, Khodaei ZS and Aliabadi M. Optimal sensor placement for maximum area coverage (MAC) for damage localization in composite structures. Smart Mater Struct 2016; 25(9): 095037.

[278] Salmanpour M, Sharif Khodaei Z and Aliabadi M. Transducer placement optimisation scheme for a delay and sum damage detection algorithm. Struct Control Hlth 2017; 24(4): e1898.

[279] Soman R, Malinowski P, Kudela P, et al. Analytical, numerical and experimental formulation of the sensor placement optimization problem for guided waves. In: 9th European workshop on structural health monitoring (EWSHM 2018), Manchester, 10–13 July 2018.

[280] Behera S, Sahoo S and Pati B. A review on optimization algorithms and application to wind energy integration to grid. Renew Sust Energ Rev 2015; 48: 214–227.

[281] Huang Y, Ludwig SA and Deng F. Sensor optimization using a genetic algorithm for structural health monitoring in harsh environments. J Civil Struct Health Monit 2016; 6(3): 509–519.

[282] Yang C, Zhang X, Huang X, et al. Optimal sensor placement for deployable antenna module health monitoring in SSPS using genetic algorithm. Acta Astronaut 2017; 140: 213–224.

[283] Lin JF, Xu YL and Law SS. Structural damage detection-oriented multi-type sensor placement with multi-objective optimization. J Sound Vib 2018; 422: 568–589.

[284] Scott M and Worden K. A bee swarm algorithm for optimizing sensor distributions for impact detection on a composite panel. Strain 2015; 51(2): 147–155.

[285] Yi TH, Li HN and Zhang XD. Health monitoring sensor placement optimization for Canton Tower using immune monkey algorithm. Struct Control Hlth 2015; 22(1): 123–138.

[286] Zhou GD, Yi TH, Zhang H, et al. Energy-aware wireless sensor placement in structural health monitoring using hybrid discrete firefly algorithm. Struct Control Hlth 2015; 22(4): 648–666.

[287] Tong K., Bakhary N., Kueh A., Yassin A.Y. M., Optimal sensor placement for mode shapes using improved simulated annealing, Smart Struct Syst 2014, 13(3), 389–406.

[288] Rashedi E., Nezamabadi-Pour H., Saryazdi S., Filter modeling using gravitational search algorithm. Eng Appl Artif Intel 2011; 24(1), 117–122.

[289] Biswas A., Mishra K., Tiwari S., Misra A. K., Physics-inspired optimization algorithms: A survey. J Optimiz 2013; 2013, 438152.

[290] Yi T.H., Li H.N., Methodology developments in sensor placement for health monitoring of civil infrastructures. Int J Distrib Sens N 2012, 8(8), 612726.

[291] Zhu, K., Gu, C., Qiu, J., Liu, W. and Fang, C., Li, B. (2016) Determining the optimal placement of sensors on a concrete arch dam using a quantum genetic algorithm, Journal of Sensors, p. 2567305.

[292] Biglar M., Gromada M., Stachowicz F., Trzepieciński, T., Optimal configuration of piezoelectric sensors and actuators for active vibration control of a plate using a genetic algorithm, Acta Mech. 2015, 226, 3451–3462.

[293] Chhabra D., Bhushan G., Chandna P., Optimal placement of piezoelectric actuators on plate structures for active vibration control via modified control matrix and singular value decomposition approach using modified heuristic genetic algorithm, Mech. Adv. Mater. Struct. 2016, 23, 272–280.

[294] Bendine K., Wankhade R.L., Optimal shape control of piezolaminated beams with different boundary condition and loading using genetic algorithm, Int. J. Adv. Struct. Eng. 2017, 9, 375–384.

[295] Samir K., Brahim B., Capozucca R., Wahab M.A., Damage detection in CFRP composite beams based on vibration analysis using proper orthogonal decomposition method with radial basis functions and cuckoo search algorithm, Compos. Struct. 2018, 187, 344–353.

[296] Ismail Z., Mustapha S., Tarhini H., Optimizing the placement of piezoelectric wafers on closed sections using a genetic algorithm – Towards application in structural health monitoring, Ultrasonics 2021, 116, 106523, https://doi.org/10.1016/j.ultras.2021.106523.

[297] Daraji, A. H., Hale, J. M. and Ye, J., New methodology for optimal placement of piezoelectric sensor/ actuator pairs for active vibration control of flexible structures', Vibration and Acoustics, Transactions of the ASME 2018, 140(1), 011015.

[298] Huang Q., Chen S., Pu H., Zhang N., Optimal piezoelectric actuators and sensors configuration for vibration suppression of aircraft framework using particle swarm algorithm, Mathematical Problems in Engineering 2017, vol. 2017, 7213125, https://doi.org/10.1155/2017/7213125.

[299] Kouider B., Polat A., Optimal position of piezoelectric actuators for active vibration reduction of beams, Applied Mathematics and Nonlinear Sciences 2020, 5(1), 385–392.

[300] Bendine K., Boukhoulda F.B., Haddag B., Nouari M., Active vibration control of composite plate with optimal placement of piezoelectric patches, Mech. Adv. Mater. Struct 2017, 26(4), 341–349.

[301] Khatir S., Belaidi I., Serra R., Wahab M.A., Khatir T., Damage detection and localization in composite beam structures based on vibration analysis, Mechanics 2015, 21, 472–479.

[302] Cantero-Chinchilla S., Beck J.L., Chiachío J., Chiachío M., Chronopoulos D., OptiSens – Convex optimization of sensor and actuator placement for ultrasonic guided-wave based structural health monitoring, SoftwareX 2021, 13, 100643, https://doi.org/10.1016/j.softx.2020.100643.

[303] Tarhini H., Itani R., Fakih M.A., Mustapha S., Optimization of piezoelectric wafer placement for structural health-monitoring applications, Intelligent Material Systems and Structures 2018, 29(19), 3758–3773.

[304] Sim SH. Estimation of flexibility matrix of beam structures using multisensor fusion. J Struct Int Maint 2016; 1(2): 60–64.

[305] Soman R, Kyriakides M, Onoufriou T, et al. Numerical evaluation of multi-metric data fusion based structural health monitoring of long span bridge structures. Struct Infrastruct E 2018; 14(6): 673–684.

[306] Xu YL, Zhang XH, Zhu S, et al. Multi-type sensor placement and response reconstruction for structural health monitoring of long-span suspension bridges. Sci Bull 2016; 61(4): 313–329.

[307] Katoch S., Chauhan S.S., Kumar, V. A., review on genetic algorithm: Past, present, and future, Multimed Tools Appl. 2021, 80, 8091–8126, https://doi.org/10.1007/s11042-020-10139-6.

[308] Kim Y.H., Kim D.H., Han J.H., and Kim C.G., Damage assessment in layered composites using spectral analysis and Lamb wave, Composites: Part B, 38, 800–809, 2007.

[309] Paget C.A., Grondel S., Levin K., and Delebarre C., Damage assessment in composites by lamb waves and wavelet coefficients, Smart Materials and Structures, 12, 393–402, 2003.

[310] Zhao X., Zhang G., Gao H., Ayhan B., Yan F., Kwan C., and Rose J.L., Active health monitoring of an aircraft wing with embedded piezoelectric sensor/actuator network: I. defect detection, localization and growth monitoring, Smart Materials and Structures, 16, 1208–1217, 2007.

[311] De Marchi, L.; Perelli, A.; Marzani, A. A signal processing approach to exploit chirp excitation in Lamb wave defect detection and localization procedures. Mech. Syst. Signal Process. 2013, 39, 20–31.

[312] Jha R., and Watkins R., Lamb wave based diagnostics of composite plates using a modified time reversal method, Proceeding of 50th AIAA/ASME/ASCE/AHS/ASC Structures, Structural Dynamics, and Materials Conference, Palm Springs, California, 4–7 May 2009.

[313] Wu, G.; Xu, C.; Du, F.; Zhu, W. A modified time reversal method for guided wave detection of bolt loosening in simulated thermal protection system panels, Complexity 2018, vol. 2018, 8210817. https://doi.org/10.1155/2018/8210817.

[314] Park, H.W.; Sohn, H.; Law, K.H.; Farrar, C.R. Time reversal active sensing for health monitoring of a composite plate. Sound Vib. 2007, 302, 50–66.

[315] Watkins, R.; Jha, R. A modified time reversal method for Lamb wave based diagnostics of composite structures. Mech. Syst. Signal Process. 2012, 31, 345–354.

[316] Liu, Z.; Yu, H.T.; Fan, J.W.; Hu, Y.A.; He, C.F.; Wu, B. Baseline-free delamination inspection in composite plates by synthesizing noncontact air-coupled Lamb wave scan method and virtual time reversal algorithm. Smart Mater. Struct. 2015, 24, 045014.

[317] Agrahari, J.K.; Kapuria, S. Effects of adhesive, host plate, transducer and excitation parameters on time reversibility of ultrasonic Lamb waves. Ultrasonics 2016, 70, 147–157.

[318] Xu, B.L.; Giurgiutiu, V. Single mode tuning effects on Lamb wave time reversal with piezoelectric wafer active sensors for structural health monitoring. Nondestruct. Eval. 2007, 26, 123–134.

[319] Zeng, L.; Lin, J.; Huang, L.P. A modified lamb wave time-reversal method for health monitoring of composite structures. Sensors 2017, 17, 955.

[320] Huang, L.P.; Zeng, L.; Lin, J.; Luo, Z. An improved time reversal method for diagnostics of composite plates using Lamb waves. Compos. Struct. 2018, 190, 10–19.

[321] Agrahari, J.K.; Kapuria, S. A refined Lamb wave time-reversal method with enhanced sensitivity for damage detection in isotropic plates. Intell. Mater. Syst. Struct. 2015, 163, 1429–1436.

[322] Mustapha, S.; Ye, L.; Dong, X.J.; Alamdari, M.M. Evaluation of barely visible indentation damage (BVID) in CF/EP sandwich composites using guided wave signals. Mech. Syst. Signal Process 2016, 76–77, 497–517.

[323] Lu Y., Wang X., Tang J., and Ding Y., Damage detection using piezoelectric transducers and the lamb wave approach: Ii, robust and quantitative decision making, Smart Materials and Structures, 17, 1–13, 2008.

[324] Ng C.T. and Veidt M., A lamb-wave-based technique for damage detection in composite laminates, Smart Materials and Structures, 18, 1–12, 2009.

[325] Su Z. and Ye L., A fast damage locating approach using digital damage fingerprints extracted from lamb wave signals. Smart Materials and Structures, 14, 1047–1054, 2005.

[326] Aitkenhead M.J., and McDonald A.J.S., A neural network face recognition system, Engineering Applications of Artificial Intelligence, 16, 167–176, 2003.

[327] Wang L. and Yuan F.G., Active damage localization technique based on energy propagation of lamb waves, Smart Structures and Systems, 3(2), 201–217, 2007.

[328] Nikles, M.; Briffod, F. Greatly extended distance pipeline monitoring using fibre optics. In Proceedings of the 24th International Conference on Offshore Mechanics and Arctic Engineering, Halkidiki, Greece, 12–17 June 2005.

[329] Inaudi, D.; Glisic, B. Distributed fibre-optic sensing for long-range monitoring of pipelines. In Proceedings of the 3rd International Conference on Structural Health Monitoring of Intelligent Infrastructure, Vancouver, BC, Canada, 13–16 November 2007.

[330] Meinert, D.; Gorny, M.; Pollmann, A.; Chen, J.; Garbi, A. Monitoring ultrasonic noise in steel pipeline. In Proceedings of the 7th International Pipeline Conference, Calgary, AB, Canada, 29 September–3 October 2008.

[331] Yan, S.; Chyan, L. Performance enhancement of BOTDR fiber optic sensor for oil and gas pipeline monitoring. Opt. Fiber Technol. 2010, 16, 100–109.

[332] Steffen, V., Jr.; Rade, D.A. Impedance-Based Structural Health Monitoring. In Dynamics of Smart Systems and Structures; Lopes Junior, V., Steffen, V. Jr., Savi, M., Eds.; Springer: Cham, Switzerland, 2016; pp. 311–328. https://doi.org/10.1007/978-3-319-29982-2_13.

[333] Wu, Z.; Zhang, H.; Yang, C. Development and performance evaluation of non-slippage optical fiber as Brillouin scattering-based distributed sensors. Struct. Health Monit. 2010, 9, 413–431, https://doi.org/10.1177/1475921710361328.

[334] Fathnejat, H.; Ahmadi-Nedushan, B.; Hosseininejad, S.; Noori, M.; Altabey, W.A. A data-driven structural damage identification approach using deep convolutional-attention-recurrent neural architecture under temperature variations. Eng. Struct. 2023, 276, 115311, https://doi.org/10.1016/j.engstruct.2022.115311.

[335] Silik, A.; Noori, M.; Ghiasi, R.; Wang, T.; Kuok, S.; Farhan, N.S.D.; Dang, J.; Wu, Z.; Altabey, W.A. Dynamic wavelet neural network model for damage features extraction and patterns recognition. Civ. Struct. Health Monit. 2023, https://doi.org/10.1007/s13349-023-00683-8.

[336] Senthilkumar, M.; Sreekanth, T.; Manikanta Reddy, S. Nondestructive health monitoring techniques for composite materials: A review. Polym. Polym. Compos. 2021, 29, 528–540.

[337] Liu, C.; Wang, X.; Noori, M.; Chang, X.; Wu, Z.; Huang, H.; Altabey, W. A., Innovative design and sensing performance of a novel large-strain sensor for prestressed FRP plates, Developments in the Built Environment, 20, 100567, November 5, 2024, https://doi.org/10.1016/j.dibe.2024.100567.

[338] Hsu, F.H. Behind Deep Blue: Building the Computer that Defeated the World Chess Champion; Princeton University Press: Princeton, NJ, USA, 2002.

[339] Ghahramani, Z. Probabilistic machine learning and artificial intelligence. Nature 2015, 521, 452–459.

[340] Müller, A.C.; Guido, S. Introduction to Machine Learning with Python: A Guide for Data Scientists; O'Reilly Media, Inc.: Newton, MA, USA, 2016.

[341] Guyon, I.; Elisseeff, A. An introduction to variable and feature selection. Mach. Learn. Res. 2003, 3, 1157–1182.

[342] Du, G.; Li, J.; Wang,W.; Jiang, C.; Song, S. Detection and characterization of stress-corrosion cracking on 304 stainless steel by electrochemical noise and acoustic emission techniques. Corros. Sci. 2011, 53, 2918–2926.

[343] Taheri, H.; Koester, L.W.; Bigelow, T.A.; Faierson, E.J.; Bond, L.J. In situ additive manufacturing process monitoring with an acoustic technique: Clustering performance evaluation using K-means algorithm. Manuf. Sci. Eng. 2019, 141, 041011.

[344] Risheh, A.; Tavakolian, P.; Melinkov, A.; Mandelis, A. Infrared computer vision in non-destructive imaging: Sharp delineation of subsurface defect boundaries in enhanced truncated correlation photothermal coherence tomography images using K-means clustering. NDT E Int. 2022, 125, 102568.

[345] Sophian, A.; Tian, G.; Fan, M. Pulsed eddy current non-destructive testing and evaluation: A review. Chin. J. Mech. Eng. 2017, 30, 500–514.

[346] Zhou, X.; Wang, H.; Hsieh, S.J.T. Thermography and K-means Clustering Methods for Anti-reflective Coating Film Inspection: Scratch and Bubble Defects. In Thermosense: Thermal Infrared Applications XXXVIII; SPIE: Bellingham, WA, USA, 2016; Volume 9861, pp. 195–204.

[347] Chen, C.; Zhou, S.; Hu, G.; Jia, J.; Floor, G.W.; Sharath, D.; Menaka, M. Lamb wave detection and localization of multiple discontinuities for plate-like structures based on DBSCAN and k-means. Mater. Eval. 2019, 77, 1439–1449.

[348] Baron, J.; Dolbey, M.; Erven, J.; Booth, D.; Murray, D. Improved pressure tube inspection in Candu reactors. Nucl. Eng. Int. 1981, 26, 45–48.

[349] Lee, H.; Lim, H.J.; Skinner, T.; Chattopadhyay, A.; Hall, A. Automated fatigue damage detection and classification technique for composite structures using Lamb waves and deep autoencoder. Mech. Syst. Signal Process. 2022, 163, 108148.

[350] Volker, C.; Kruschwitz, S.; Boller, C.; Wiggenhauser, H. Feasibility study on adapting a machine learning based multi-sensor data fusion approach for honeycomb detection in concrete. In NDE/NDT for Highways & Bridges: SMT 2016; The American Society of Nondestructive Testing: Columbus, OH, USA, 2016; pp. 144–148.

[351] Selim, H.; Delgado Prieto, M.; Trull, J.; Romeral, L.; Cojocaru, C. Laser ultrasound inspection based on wavelet transform and data clustering for defect estimation in metallic samples. Sensors 2019, 19, 573.

[352] Li, T.J.; Chen, C.C.; Liu, J.J.; Shao, G.F.; Chan, C.C.K. A novel THz differential spectral clustering recognition method based on t-SNE. Discret. Dyn. Nat. Soc. 2020, 2020, 6787608.

[353] Li, X.; Shen, T. Nondestructive testing system design for biological product based on vibration signal analysis of acceleration sensor. J. Vibroeng. 2017, 19, 2164–2173.

[354] Sevillano, E.; Sun, R.; Gil, A.; Perera, R. Interfacial crack-induced debonding identification in FRP-strengthened RC beams from PZT signatures using hierarchical clustering analysis. Compos. Part B Eng. 2016, 87, 322–335.

[355] Salazar, A.; Igual, J.; Vergara, L. Agglomerative clustering of defects in ultrasonic non-destructive testing using hierarchical mixtures of independent component analyzers. In Proceedings of the 2014 International Joint Conference on Neural Networks (IJCNN), Beijing, China, 6–11 July 2014; pp. 2042–2049.

[356] Na, W.S. Low cost technique for detecting adhesive debonding damage of glass epoxy composite plate using an impedance based non-destructive testing method. Compos. Struct. 2018, 189, 99–106.

[357] Shao, J.; Shi, H.; Du, D.; Wang, L.; Cao, H. Automatic weld defect detection in real-time X-ray images based on support vector machine. In Proceedings of the 4th International Congress on Image and Signal Processing, Shanghai, China, 15–17 October 2011; Volume 4, pp. 1842–1846.

[358] Lee, L.H.; Rajkumar, R.; Lo, L.H.; Wan, C.H.; Isa, D. Oil and gas pipeline failure prediction system using long range ultrasonic transducers and Euclidean-support vector machines classification approach. Expert Syst. Appl. 2013, 40, 1925–1934.

[359] Tian, F.; Tan, F.; Li, H. An rapid nondestructive testing method for distinguishing rice producing areas based on Raman spectroscopy and support vector machine. Vib. Spectrosc. 2020, 107, 103017.

[360] Moomen, A.; Ali, A.; Ramahi, O.M. Reducing sweeping frequencies in microwave NDT employing machine learning feature selection,. Sensors 2016 16, 559.

[361] Saleem, M.; Gutierrez, H. Using artificial neural network and non-destructive test for crack detection in concrete surrounding the embedded steel reinforcement. Struct. Concr. 2021, 22, 2849–2867.

[362] Saeed, N.; Omar, M.A.; Abdulrahman, Y. A neural network approach for quantifying defects depth, for nondestructive testing thermograms. Infrared Phys. Technol. 2018, 94, 55–64.

[363] Kwon, D.; Kim, H.; Kim, J.; Suh, S.C.; Kim, I.; Kim, K.J. A survey of deep learning-based network anomaly detection. Clust. Comput. 2019, 22, 949–961.

[364] Goodfellow, I.; Bengio, Y.; Courville, A. Deep Learning; MIT Press: Cambridge, MA, USA, 2016.

[365] Scherer, D.; Müller, A.; Behnke, S. Evaluation of pooling operations in convolutional architectures for object recognition. In Proceedings of the International Conference on Artificial Neural Networks, Thessaloniki, Greece, 15–18 September 2010; Springer: Berlin/Heidelberg, Germany, 2010; pp. 92–101.

[366] Kiranyaz, S.; Waris, M.A.; Ahmad, I.; Hamila, R.; Gabbouj, M. Face segmentation in thumbnail images by data-adaptive convolutional segmentation networks. In Proceedings of the 2016 IEEE International Conference on Image Processing (ICIP), Phoenix, AZ, USA, 25–28 September 2016; IEEE: Piscataway, NJ, USA, 2016; pp. 2306–2310.

[367] Kiranyaz, S.; Ince, T.; Gabbouj, M. Real-time patient-specific ECG classification by 1-D convolutional neural networks. IEEE Trans. Biomed. Eng. 2015, 63, 664–675.

[368] Dreyfus, H.L, What Computers Can't Do – The Limits of Artificial Intelligence. New York: Harper and Row, 1979.

[369] Fukushima, K., A neural network for visual pattern recognition. Computer 21, 65–75, 1988.

[370] Baxt WG., Use of an artificial neural network for the diagnosis of myocardial infarction. Annals of Internal Medicine, 115(11), 843–848, 1991.

[371] Astion M.L., Wener M.H., Thomas R.G., Hunder G.G. and Bloch D.A., Application of neural networks to the classification of giant cell arteritis. Arthritis and Rheumatism, 37(5), 760–770, 1994.

[372] Wong F.S., A 3D neural network for business forecasting. Proceedings of the 24th Annual Hawaii International Conference on Systems Sciences, 4, 113–123, 1991.

[373] Widrow B., Rumelhart D.E. and Lehr M.A., Neural networks: Applications in industry, business and science. Communications of the ACM 37(3), 93–105, 1994.

[374] Jensen H., Using neural networks for credit scoring, Managerial Finance, 15–26, 1992.

[375] Refenes A.P., Neural Networks in the Capital Markets, Wiley, 1995.

[376] Bose, N.K. and Liang P., Neural Network Fundamentals with Graphs, Algorithms, and Applications. McGraw-Hill, 1996.

[377] Hagan M.T., Demuth, H.B. and De Jesús O., An introduction to the use of neural networks in control systems. International Journal of Robust and Nonlinear Control, 12(11), 959–985, 2002.

[378] Howard S., Neural networks in electrical engineering. Proceedings of the ASEE New England Section 2006 Annual Conference (Session 1A – Electrical & Computer Engineering).

[379] Pirdashti M., Curteanu S., Kamangar M.H. and Khatami M.A., Artificial neural networks: Applications in chemical engineering, Reviews in Chemical Engineering 24(9), 205–239, 2013.

[380] Zhao, Y., Noori, M., Altabey, W.A., Ghiasi, R. and Zhishen, W. (2018). Deep learning-based damage, load and support identification for a composite pipeline by extracting modal macro strains from dynamic excitations. Applied Sciences, 8(12), 2564, https://doi.org/10.3390/app8122564.

[381] Kost, A., Altabey, W. A., Noori, M. and Awad, T. (2019). Applying neural networks for tire pressure monitoring systems. Structural Durability and Health Monitoring, 13(3), 247–266, http://dx.doi.org/10.32604/sdhm.2019.07025.

[382] Wang, T., Altabey, W. A., Noori, M. and Ghiasi, R. (2020). A deep learning based approach for response prediction of beam-like structures. In Press, Structural Durability and Health Monitoring, http://www.tspsubmission.com/index.php/sdhm.

[383] Kamble B.C., Speech recognition using artificial neural network– a review, Int'l Journal of Computing, Communications & Instrumentation Engg. (IJCCIE), 3(1), 1–4, 2016.

[384] Wouter G., Georgi T. and Valeri M., Neural network used for speech recognition, Journal of Automatic Control, University of Belgrade, 20, 1–7, 2010. http://dx.doi.org/10.2298/JAC1001001G

[385] Yashwanth H., Harish M. and Suman D., Automatic Speech recognition Using Audio Visual Cues, IEEE India Annual Conference; pp. 166–169, 2004.

[386] Ozyilmaz, L. and Yildirim T., Artificial neural networks for diagnosis of hepatitis disease, Proceedings of the International Joint Conference on Neural Networks, 1, 586–589, 20–24, 2003.

[387] Aleksander I. and Morton H., An Introduction to Neural Computing. Int Thomson Comput Press, London, 1995.

[388] Aitkenhead M.J. and McDonald A.J.S., A neural network face recognition system, Engineering Applications of Artificial Intelligence, 2003.

[389] Alippi C., Real-time analysis of ships in radar images with neural networks, Pattern Recognition, 1995.

[390] Anwar M. and Farzin D., Neural networks for the classification of image texture, Engineering Applications of Artificial Intelligence, Neural Networks, 1994.

[391] Hanagud S., and Luo H., Damage Detection and Health Monitoring Based on Structural Dynamics, Structural Health Monitoring, Current Status and Perspectives, Stanford University, Palo Alto, California, USA, 715–726, 1997.

[392] Luo H., and Hanagud S., Dynamic learning rate neural network training and composite structural damage detection, AIAA Journal, 35(9), 1522–1527, 1997.

[393] Krawczuk M., Ostachowicz W., and Kawiecki G., Detection of delaminations in cantilevered beams using soft computing methods, Proceedings of the European COST F3 Conference on System Identification and Structural Health Monitoring, Madrid, Spain, 243–252, 2000.

[394] Messina A., Jones I.A., and Williams E.J., Damage detection and localization using natural frequency changes, Proceedings of the 1st Conference on Identification, Cambridge, England, UK, 67–76, 1992.

[395] Hatem T.M., Foutouh M.N.A., and Negm H.M., Application of genetic algorithms and neural networks to health monitoring of composite structures, Proceedings of the 2nd European Workshop on Structural Health Monitoring, Forum am Deutschen Museum, Munich, Germany, 616–623, 2004.

[396] Zheng D., Low Velocity Impact Analysis of Composite Laminated Plates, PhD Thesis, Graduate Faculty of the University of Akron, USA, 2007.

[397] Haj-Ali R.M., Pecknold D.A., Ghaboussi J., and Voyiadjis G.Z., Simulated micromechanical models using artificial neural networks, Engineering Mechanics, 127 (7), 730–738, 2001.

[398] Chakraborty D., Artificial neural network based delamination prediction in laminated composites, Materials and Design, 26, 1–7, 2005.

[399] Roseiro L., Ramos U., and Leal R., Neural networks in damage detection of composite laminated plates, Proceedings of the 6th WSEAS International Conference on Neural Networks, Lisbon, Portugal, 115–119, June 16–18, 2005.

[400] Karnik S.R., Gaitonde V.N., Rubio J.C., Correia E., Abrão A.M. and Davim J.P., Delamination analysis in high speed drilling of carbon fibre reinforced plastics (CFRP) using artificial neural network model, Materials and Design, 29 (9), 1768–1776, 2008.

[401] Saadat, S., Noori, M., and Buckner, G., T. Furukawa and Y. Suzuki Structural health monitoring and damage detection using an intelligent parameter varying (IPV) technique, International Journal of Nonlinear Mechanics, Volume 39, No. 10, pp. 1687–1697, 2004.

[402] Saadat, S., Buckner, G., Noori, M., Structural system identification and damage detection using the IPV technique: An experimental study, Structural Health Monitoring; vol. 6:3, pp. 231–243, 2007

[403] Hen Y. and Hwang JN., Handbook of Neural Network Signal Processing, CRC press, Boca Raton, 2001.

[404] Zhao Y., Noori M., Altabey W.A. and Lu N., Reliability evaluation of a laminate composite plate under distributed pressure using a hybrid response surface method, International Journal of Reliability, Quality and Safety Engineering, 24(2), 2017.

[405] Meruane, V.; Aichele, D.; Ruiz, R.; López Droguett, E. A deep learning framework for damage assessment of composite sandwich structures. Shock Vib. 2021, 2021, 1483594.

[406] Fotouhi, S.; Pashmforoush, F.; Bodaghi, M.; Fotouhi, M. Autonomous damage recognition in visual inspection of laminated composite structures using deep learning. Compos. Struct. 2021, 268, 113960.

[407] Seventekidis, P.; Giagopoulos, D. A combined finite element and hierarchical Deep learning approach for structural health monitoring: Test on a pin-joint composite truss structure. Mech. Syst. Signal Process. 2021, 157, 107735.

[408] Tran-Ngoc, H.; Khatir, S.; Ho-Khac, H.; De Roeck, G.; Bui-Tien, T.; Wahab, M.A. Efficient Artificial neural networks based on a hybrid metaheuristic optimization algorithm for damage detection in laminated composite structures. Compos. Struct. 2021, 262, 113339.

[409] Zenzen, R.; Khatir, S.; Belaidi, I.; Le Thanh, C.; Wahab, M.A. A modified transmissibility indicator and artificial neural network for damage identification and quantification in laminated composite structures. Compos. Struct. 2020, 248, 112497.

[410] Muir, C.; Swaminathan, B.; Fields, K.; Almansour, A.; Sevener, K.; Smith, C.; Presby, M.; Kiser, J.; Pollock, T.; Daly, S. A machine learning framework for damage mechanism identification from acoustic emissions in unidirectional SiC/SiC composites. NPJ Comput. Mater. 2021, 7, 146.

[411] Fu Jin-Gang, Zhu Dong-Mei, Zhou Wan-Cheng, Luo Fa, Anisotropic dielectric properties of short carbon fiber composites, Inorganic Materials, 27(11), 1223–1227 (2012), https://doi.org/10.3724/SP.J.1077.2012.12364.

[412] N. Angelidis, N. Khemiri and P. E. Irving, Experimental and finite element study of the electrical potential technique for damage detection in CFRP laminates, Smart Materials and Structures, 14 (1), 147.

[413] Yoshiyasu Hirano, Takuya Yamane, Akira Todoroki, Through-thickness electrical conductivity of toughened CFRP laminate, Composites Science and Technology, 122 (18), 2016, 67–72.

[414] F. Ellyin and R. Maser, Environmental effects on the mechanical properties of glass-fiber epoxy composite tubular specimens, Composites Science and Technology, 64, 1863–1874 (2004).

[415] GS. Springer, Environmental effects, Ch. 1, PA: Technomic Pub. Co. Springer GS, Environmental effects on composite materials, (eds.3) (1981).

[416] Jr. EL. McKague, J.D. Reynolds and J.E. Halkias, Moisture diffusion in fiber reinforced plastics, Eng Mater Technol., 98, 92–95 (1976).

[417] J.R.M. d'Almeida, R. C. de Almeida and W. R. De Lima, Effect of water absorption of the mechanical behavior of fiberglass pipes used for offshore service waters, Composite Structures, 83, 221–225 (2008).

[418] Davis, J.; Goadrjch, M. The relationship between precision-recall and ROC curves. In Proceeding of the 23rd International Conference on Machine Learning, Pittsburgh, PA, USA, 25-29 June 2006; ACM: New York, NY, USA, 2006; 233–240.

[419] Fawcetr, T. An introduction to ROC analysis. Pattern Recognit. Lett. 2006, 27, 861–874.

[420] Dogan T., Prediction of Composite Vessels Under Various Loadings, Master Thesis, 2006, Dokuz Eylul University, Department of Mechanical Engineering, Izmir, Turkey.

[421] Azzi V. D. and Tsai S. W., Anisotropic strength of composites – Investigation aimed at developing a theory applicable to laminated as well as unidirectional composites, employing simple material properties derived from unidirectional specimens alone, Journal of Experimental Mechanics, 1965, 5(9), pp. 283–288.

[422] Ian, M., (2012), Ward, John Sweeney, Mechanical Properties of Solid Polymers, John Wiley & Sons.

[423] Drabousky, D. P. (2009), Prony Series Representation and Interconversion of Viscoelastic Material Functions of Equine Cortical Bone, Master thesis, Department of Mechanical and Aerospace Engineering, Case Western Reserve University.

[424] Rafiee, R., Ghorbanhosseini, A., (2020) Developing a micro-macromechanical approach for evaluating long-term creep in composite cylinders, Thin–Walled Structures, 151, 106714, https://doi.org/10.1016/j.tws.2020.106714.

[425] Rafiee, R., Ghorbanhosseini, A., (2020) Analyzing the long-term creep behavior of composite pipes: Developing an alternative scenario of short-term multi-stage loading test, Composite Structures, 254, 112868, https://doi.org/10.1016/j.compstruct.2020.112868.

[426] Noori, M., Altabey, W. A. (2022), Hysteresis in engineering systems. Appl. Sci., 12(19), 9428. https://doi.org/10.3390/app12199428.

[427] Altabey, W.A.; Noori, M.; Wu, Z.; Al-Moghazy, M.A.; Kouritem, S.A. Studying acoustic behavior of BFRP laminated composite in dual-chamber muffler application using deep learning algorithm. Materials 2022, 15, 8071. https://doi.org/10.3390/ma15228071.

[428] Wang, T.; Li, H.; Noori, M.; Ghiasi, R.; Kuok, S.; Altabey, W.A. Seismic response prediction of structures based on Runge-Kutta recurrent neural network with prior knowledge. Eng. Struct. 2023, 279, 115576, https://doi.org/10.1016/j.engstruct.2022.115576.

[429] Wang, H.; Feng, S.; Gong, X.; Guo, Y.; Xiang, P.; Fang, Y.; Li, Q. Dynamic performance detection of CFRP composite pipes based on quasi-distributed optical fiber sensing techniques. Front. Mater. 2021, 8, 683374, http://dx.doi.org/10.3389/fmats.2021.683374.

[430] Altabey, W.A. The damage identification in laminated composite plate under fatigue load through wavelet packet energy curvature difference method. Compos. Part C Open Access 2022, 9, 100304. https://doi.org/10.1016/j.jcomc.2022.100304.

[431] Sheng, M.; Wang, M.; Sun, J. The Basis of Noise and Vibration Control Technology; Science Press: Beijing, China, 2001.

[432] Selamet, A.; Ji, Z.L. Acoustic attenuation performance of circular expansion chambers with extended inlet/outlet. Sound Vib. 1999, 223, 197–212.

[433] Altabey, W.A. Applying deep learning and wavelet transform for predicting the vibration behavior in variable thickness skew composite plates with intermediate elastic support. J. Vibroeng. 2021, 23, 770–783. https://doi.org/10.21595/jve.2020.21480.

[434] Li, Z.; Noori, M.; Wan, C.; Yu, B.; Wang, B.; Altabey, W.A. A deep learning-based approach for the identification of a multi-parameter BWBN model. Appl. Sci. 2022, 12, 9440. https://doi.org/10.3390/app12199440.

[435] Ramachandran, P.; Zoph, B.; Quoc, V.L. Searching for Activation Functions. arXiv 2017, arXiv:1710.05941. https://doi.org/10.48550/arXiv.1710.05941.

[436] Altabey, W.A.; Noori, M.; Wang, T.; Ghiasi, R.; Kuok, S.-C.; Wu, Z. Deep learning-based crack identification for steel pipelines by extracting features from 3d shadow modeling. Appl. Sci. 2021, 11, 6063. https://doi.org/10.3390/app11136063.

[437] Olah, C. Understanding LSTM Networks. 2015. Available online: https://colah.github.io/posts/2015-08-Understanding-LSTMs/ (accessed on 5 Oct. 2022).

[438] Wang, T.; Li, H.; Noori, M.; Ghiasi, R.; Kuok, S.-C.; Altabey, W.A. Probabilistic seismic response prediction of three-dimensional structures based on Bayesian convolutional neural network. Sensors 2022, 22, 3775. https://doi.org/10.3390/s22103775.

[439] Zachary, C.L.; John, B.; Charles, E. A critical review of recurrent neural networks for sequence learning. arXiv 2015, arXiv:1506.00019. https://doi.org/10.48550/arXiv.1506.00019

[440] Hutter, F.; Hoos, H.H.; Leyton-Brown, K. Sequential Model-based Optimization for General Algorithm Configuration. In Learning and Intelligent Optimization; Springer: Berlin/Heidelberg, Germany, 2011; pp. 507–523.

[441] Diederik, P.K.; Jimmy, B. Adam: A Method for Stochastic Optimization. arXiv 2014, arXiv:1412.6980. https://doi.org/10.48550/arXiv.1412.6980.

[442] Martinez-Cantin, R. Bayesopt: A Bayesian optimization library for nonlinear optimization, experimental design and bandits. Mach. Learn. Res. 2014, 15, 3735–3739.

[443] Mohebian, P.; Aval, S.B.B.; Noori, M.; Lu, N.; Altabey, W.A. Visible particle series search algorithm and its application in structural damage identification. Sensors 2022, 22, 1275. https://doi.org/10.3390/s22031275.

[444] Bergstra, J., Bardenet, R., Bengio, Y., Kégl, B. Algorithms for hyper-parameter optimization. In Proceedings of the Advances in Neural Information Processing Systems, *Granada, Spain, 12–15 December* 2011; pp. 2546–2554.

[445] Bergstra, J.; Yamins, D.; Cox, D. Making a science of model search: Hyperparameter optimization in hundreds of dimensions for vision architectures. In Proceedings of the 30th International Conference on Machine Learning, Atlanta, GA, USA, 16–21 June 2013; pp. 115–123.

[446] Liu, H.; Lin, J.; Hua, R.; Dong, L. Structural optimization of a muffler for a marine pumping system based on numerical calculation. J. Mar. Sci. Eng. 2022, 10, 937. https://doi.org/10.3390/jmse10070937.

Index

https://doi.org/10.1515/9783112213094-010

www.ingramcontent.com/pod-product-compliance
Lightning Source LLC
Chambersburg PA
CBHW061402210326
41598CB00035B/6066